The Saga of a Braveheart

Lt Col Ajit V Bhandarkar

Shaurya Chakra

The Saga of a Braveheart
Lt Col Ajit V Bhandarkar
Shaurya Chakra

Shakunthala Ajit Bhandarkar

Vij Books India Pvt Ltd

New Delhi (India)

Published by

Vij Books India Pvt Ltd
(Publishers, Distributors & Importers)
2/19, Ansari Road
Delhi – 110 002
Phones: 91-11-43596460, 91-11-47340674
Mobile: 98110 94883
e-mail: contact@vijpublishing.com
www.vijbooks.in

ISBN: 978-93-90917-12-9 (Hardback)
ISBN: 978-93-90917-13-6 (Paperback)
ISBN: 978-93-90917-14-3 (ebook)

Front Cover:
Cadet Sgt Ajit V Bhandarkar in his Blue Patrol,@NDA 1980. Designed by Karan Rustagi.
Back Cover: Watermark, The War memorial , IMA, Dehradun.

Dedicated to

Ajit's and my parents
for encouraging us in all our endeavours
and giving us the right values in life.

CONTENTS

Foreword		ix
Reviews		xii
Author's Note		xiv
Acknowledgement		xvii
Glossary		xxi
Chapter 1	Introduction: The Birth of a Braveheart	1
Chapter 2	Ajeet hai Abheet hai - Schooling at Sainik School Bijapur	5
Chapter 3	Seva Parmo Dharma: NDA Days	19
Chapter 4	Veerta aur Vivek – IMA: The Decision to Join the Madras Regiment	28
Chapter 5	Swadharme Nidhanam Shreyah - The Madras Regiment	36
Chapter 6	18 MADRAS (Mysore) in Gandhinagar	42
Chapter 7	Serve with Honour – Officers Training Academy: An Instructor Who Touched the Hearts of His GC	63
Chapter 8	Sulagne Savadhana – Marriage in Chennai	70
Chapter 9	Unit Life in Ferozepur	75
Chapter 10	Yudham Pragya- Defence Service Staff College, Wellington	80

Chapter 11   Brigade Major at Nathu La   87

Chapter 12   MS Branch - Army Headquarters: South Block,
             New Delhi   93

Chapter 13   Yudha Kritya Nischaya: The Army War College;
             Where the Tigers Earned Their Stripes: Ajit's
             Professional Alma Mater   98

Chapter 14   Dridhata aur Virata: 25 RR   103

Chapter 15   Shaurya Chakra: Gallantry Award   115

Chapter 16   Chip off the Old Block: Both Boys Join the
             Armed Forces   120

FOREWORD

2/Lt Ajit Bhandarkar was commissioned into 18th Battalion, The Madras Regiment (Mysore) on 19th December 1981. As a Lieutenant Colonel, Second-in-Command of his battalion, Ajit volunteered to lead a Quick Reaction Team to locate and eliminate a few Pakistani terrorists who had crossed the Line of Actual Control into Surankot area, near Poonch (Jammu and Kashmir) on 30th October 1999. When the encounter took place, he shot down three terrorists and then took a terrorist's shot on his forehead and was killed in action. For his dedication to duty, leadership from the front, personal bravery and supreme sacrifice for the nation, he was awarded the *Shaurya Chakra (SC)*.

We have read many stories and biographies of our brave soldiers and war heroes. The focus is on the braveheart's military upbringing, dedication, camaraderie and valour. Most of these narrations are authored by journalists, researchers or professional writers. In rare cases, one may also find a story co-authored with a family member.

The Saga of a Braveheart: Lt Col Ajit V Bhandarkar Shaurya Chakra is different; I would say unique in some sense. It is written by a *Veer Nari*: Mrs Shakunthala Bhandarkar, wife of late Lieutenant Colonel Ajit Bhandarkar, SC. It is a rare glimpse into the heart of a brave soldier, his family and children. Anecdotes penned by Mrs Shakunthala Bhandarkar in her very simple language bring alive the love for her husband, dedication to the family, and vividly the experiences which a *Fauji* wife goes through with her husband, and then unfortunately and most unexpectedly, without him.

Mrs Shakunthala Bhandarkar, like all *Veer Naris,* has borne the pain of her very personal loss with quiet dignity, courage and fortitude. She has brought up her two young sons single-handedly and inculcated the social and military values which are dear to her late husband and her. I am delighted and proud to say that both her sons are now commissioned officers in the armed forces. There can be no better proof of her love for the armed forces and their service to the nation.

This book takes you through Ajit's life; introduces you to his family and also gives glimpses of a military school (Sainik School, Bijapur), some important armed forces institutions and his renowned unit, 18th Battalion, The Madras Regiment (Mysore).

When Mrs Shakunthala Bhandarkar decided to write this book, she told me, "While the *Fauji* crowd is aware of the trials and tribulations our officers and their families go through, our civilian friends are unaware of the various institutions in the armed forces. They are also oblivious to the challenges we face. My objective to write this biography is to share the life of a braveheart with the younger generation and bring about awareness about the armed forces as a profession, which the youth should covet."

It is a privilege for me to write a Foreword to this book, and I salute late Lieutenant Colonel Ajit Bhandarkar, SC and also Mrs Shakunthala Bhandarkar's own exemplary courage, dedication and achievements.

Date: 13 Apr 2021   General V P Malik,
                                       PVSM, AVSM (Retired)
                              Former Chief of the Army Staff, India

Note: Gen. V P Malik, PVSM, AVSM (Retd) was the Chief of Army Staff during Operation Vijay (The Kargil War), the Op Faizalabad led by Lt Col A Bhandarkar and was also the Commandant of the DSSC while then Maj Ajit Bhandarkar was doing the DSSC course (49).

Brig. (Dr.) B. D. Mishra (Retd.)
Governor,
Arunachal Pradesh

RAJ BHAVAN
ITANAGAR - 791 111

PROEM

With a sense of privilege, I am penning these lines to reminisce the persona, patriotism and prowess of my Regimental Officer Lt. Col. Ajit Bhandarkar, Shaurya Chakra (Posthumous). Lt. Col. Ajit was a loveable lad, a confident School Student Captain and a brave Infantry Officer. During his 18 years of Commissioned service he had won the trust of his soldiers, praise of his colleagues and appreciation of his superiors. He was a true Field Commander and led his troops from the front.

On the day he made the supreme sacrifice for the unity and integrity of the Country, he was leading an operation against the Pakistan aided and armed terrorists, holed up in Faizalabad in Suran Kot Tehsil of Poonch District in Jammu and Kashmir. During the operation, he was aggressive, audacious and demonstrably plucky to neutralize the terrorists. Determined to attain his mission, he remained unconcerned about his personal safety till he fell to the perfidious bullets of the terrorists. For his bravery in the face of enemy, the Nation decorated Lt Col Ajit with Shaurya Chakra. In his departure, the Indian Army lost a gallant fighter and the Nation a noble citizen.

Over two decades after the martyrdom of Lt. Col. Ajit Bhandarkar, Shaurya Chakra (Posthumous), Mrs. Shakuntala Bhandarkar, with strong will and sustained effort, has written the Biography of her husband. It presents an authentic version of the last battle fought by Lt. Col. Ajit Bhandarkar.

Mrs. Shakuntala Bhandarkar deserves compliments for her courage in braving the difficulties and challenges which eclipse a Soldier's wife after her husband's sudden departure. At the time of Lt. Col. Bhandarkar's martyrdom their two sons Nirbhay and Akshay were respectively 7 years and 5 years old. She groomed them creditably well and prepared them to follow the gracious footsteps of their father. Nirbhay joined the Indian Army and Akshay, the Indian Navy.

Mrs. Bhandarkar is a **VEER NARI** in true sense of the term.

I am sanguine, this Biography will serve as a beckon for our youth to emulate the Military career choice of Major Nirbhay Bhandarkar and Lieutenant Akshay Bhandarkar. Equally importantly, it will surely serve as an abiding text of fortitude and perseverance for single mother parents to motivate them to bring up their children with hope and grit.

Beyond doubt, the profession of Arms in the Indian Armed Forces is the best form of **NATIONALISM**.

Brig. (Dr.) B.D. Mishra (Retd.)

REVIEWS

"Memories of good times spent together at NDA flooded back as I read this poignant story of a courageous soldier, an upright officer and a gentleman. Rising above the call of duty, leading from the front and unmindful of personal danger, Lt Col Ajit Bhandarkar epitomises all that makes the Indian army among the best in the world today."

Air Vice Marshal (Dr) Arjun Subramaniam, AVSM (Retd),
Author of *India's Wars: A Military History 1947-1971* and *Full Spectrum: India's Wars 1972-2020*.

"Though I had not known Ajit personally but after taking over command of 25 RR from Col Ramachandran, VSM, I learnt about his heroic feat and learnt about his martyrdom. Later when I began to interact with the locals of Draba, the location of the Battalion Headquarters, I felt proud when everybody talked of Ajit not only as a very fine gentleman but a brave soldier as well. The chord he had struck with the locals speaks volumes of his humanitarian qualities. Ajit has done proud not only the RR or Madras Regiment but the entire nation living up to the idea of India when militancy was at its peak in the area. My salute to him and his valour and to his devoted wife who not only offered her two sons to the Armed forces but herself has left no stone unturned to keep his legacy alive."

Brig Anil Gupta (Retd)

Former CO 25 RR

"The author, Mrs Shakunthala A Bhandarkar, has done a very praiseworthy job in writing about the life of her dear husband, Lt Col Ajit V Bhandarkar, SC. This book which reads very smoothly, brings out the leadership and humane qualities of Ajit. The travails of the tough and unforgiving Army life for an officer and his family have been very well brought out. When I took over the battalion in Gandhinagar Ajit was the Adjutant. In the short period we served together, I found Ajit to be a well-groomed young officer with sterling qualities of the head, heart and professional. He was very diligent and thorough in his work. Shortly after I took over command, the battalion moved to a field area in Jammu and Kashmir in 1986. Ajit assisted me in the planning and execution of the move and induction of the battalion into Battalion Defended Area, along the Line of Control and in his characteristic style did a very magnificent job at a very young service."

"Only brave soldiers are killed in action. Brave commanders lead from the front and Ajit like a truly brave officer led his troops right from the front."

Lt Gen Arun Kumar Chopra PVSM, AVSM (Retd)
Veteran 1971 War and 1999 Kargil War
Former Commandant of NDA
Former Col of the Madras Regiment
Former CO 18 MADRAS (Mysore)

Author's Note

Dear Readers,

On 31st December 2020, we celebrated Ajit Bhandarkar's 60th birth anniversary. He has been a guiding light to all of us. His mother still remembers him fondly as a compassionate son who would sit with her and have a chat. When on leave, he would do the chores of not only bringing her groceries but also help her deep clean the whole house.

Ajit's demise was a shocker to all of us; we lost a wonderful human being who was always ready to help others. *"Manav"* as his school mates called him for his compassion and understanding.

Immediately after Ajit's martyrdom on Saturday, the 30th October 1999, I decided to move to Bangalore, where all our family stayed. So here I was, after the kids' Academic session, April 2000 in Bangalore, coping with the loss of my life partner. I focused my energy and time in bringing up my darling sons Nirbhay and Akshay who were only five and seven years old. I knew that I had a very long way to go, and it was going to be tough; nevertheless, I accepted it. Within a few months, I got a job in Army Public School as a primary school teacher after clearing the exams and thereafter, there was no looking back.

Also let me confess, I neither had the mental strength to discuss about Ajit nor even talk about him to others. But then, in the eyes of my family members and friends, they all felt I was strong and brave, and I hid my pain from the world. I have followed this gospel truth: 'Smile, the world smiles with you and weep, you weep alone'. It has taken me long to heal and muster courage to write Ajit's biography.

While writing the book, I have frequently called Ajit's course mates and his school mates for details about their NDA days and school-days, respectively. The 18 MADRAS (Mysore), Ajit's parent unit officers, and 25 Rashtriya Rifles, where he worked under deputation and his last posting, have always been by my side, and their constant support has enabled me to accomplish the task of writing this biography.

My heart-felt gratitude to Lt Gen PG Kamath, PVSM, AVSM, YSM, SM (Retd) for guiding me and helping me with his valuable inputs. Last but not the least, Brig Pradip Vij, of Vij Publication, for accepting the book for publishing, the minute I told him about my endeavour to pen down Ajit's journey of life.

I profusely thank honourable Governor, Brig. (Dr) B D Mishra (Retd) and Gen V P Malik, PVSM, AVSM, (Retd) for taking their valuable time to read the book and give the Foreword for the book.

As a novice in writing, I have used simple language to narrate the life of a great human being. I also wanted to highlight the various defence institutions through which Ajit got trained and could master the art of leadership. I have tried my best to showcase not only Ajit's life but also give a peek into the training phase and army lifestyle, for the benefit of all my civilian friends and the youth who want to pursue a career in the Defence Forces.

The objective of writing this book is to share the life of Ajit with the whole world, the ones who are keen to know about him and also, I hope you as readers, understand that every soldier who takes the oath for defending and serving his country stands by it. No matter what, even at the cost of his life.

Ajit would never share his professional life experiences and lessons with me and therefore, I was totally ignorant of many of the challenges he had to face in his service career. I would learn of them only through other ladies of the unit and other junior officers, and sometimes it would be Ajit's buddy who would give me a first-hand report of any encounter or operations executed by Ajit.

So, to understand the last operation (Faizalabad in J&K) Ajit took part in, I had to talk to his buddy during the operation and others who witnessed it. It was during the candid confessions made by almost all the jawans, that I fully understood the camaraderie between the troops and the officer. He loved his men and cared for them like a father.

No words will suffice to express my feelings for him and the vacuum it has created in our life. It was at the behest of my son Akshay, who encouraged me to pen down my sentiments and respect for Ajit—the amazing friend, the devoted son, the loving husband and father, and the serious soldier with a heart of gold, who sacrificed his life for the country.

Life is a journey and we cannot change destiny. Ajit was destined to die on the battlefield with his boots on, which every soldier dreams of. So, I am happy in a way that he has achieved his dream and lived his life to the full as a frontline soldier.

सुखदुःखे समे कृत्वा लाभालाभौ जयाजयौ।
ततो युद्धाय युज्यस्व नैवं पापमवाप्स्यसि॥

Fight for the sake of duty, treating alike happiness and distress, loss and gain, victory and defeat.

Fulfilling your responsibility in this way, you will never incur sin.

|Bhagavad Gita 2.38|

Happy Reading.

Shakunthala

shakunthala@ajitbhandarkar.com

ACKNOWLEDGMENT

The school mates of Sainik School Bijapur, Karnataka, starting with Lt Gen PG Kamath, PVSM, AVSM, YSM, SM (Retd), who has been my main mentor in the process of writing this book, Maj Gen Arjun Muthanna (Retd), Maj Gen K S Kumbar, VSM (Retd), Brig SB Sajjan (Retd), Brig Mohan Pattar (Retd), Col Aniruddha Narasimha Gudi (Retd), Col Som Neeraj Roy (Retd), Mr Gopal Kulkarni, Mr Shivaprasad Khened and others for giving valuable inputs about the school life, recalling academy days and some interesting anecdotes of Ajit. Till date including the present Principal of the school Capt. Vinayak Tiwari (Retd) and his team have given me all the possible help whenever I needed.

The 59th NDA and 69th IMA course mates, the Golf Squadron mates and other NDA fellow mates for giving me almost all big and small inputs about Ajit's life during the academy days. Starting from Lt Gen Rajeev Sabherwal, PVSM, AVSM, VSM (Retd), Lt Gen YVK Mohan, PVSM, AVSM, SM, VSM (Retd), Maj Gen Anuj Mathur, VSM (Retd), Brig Ved Prakash (Retd), Brig Vikas Puri (Retd), Col G K Rao (Retd), Col Rajesh Tiwari (Retd), Col Pradeep Katoch (Retd) , Col NN Singh (Retd), Col A S Chandoke (Retd), Col Vikram Tiwathia (Retd), Sqd Ldr Ashwani Puri, (Retd), Col Nishant Kaura (Retd), Col Ram Athawale (Retd), Col Pankaj Goel (Retd), Capt (Dr) Probal Kumar Ghosh (IN) (Retd), Col Rajesh Kapur (Retd) and others for all their assistance and help by way of sharing anecdotes, photographs and giving ideas to improvise the book. Special thanks to the authors in the 59th Course (NDA), Air Vice Marshal (Dr) Arjun Subramaniam, AVSM (Retd), and Mukul Deva, for guiding me and giving their valuable suggestions in the making of the book.

Col OLV Naresh, SM, CO of 18 MADRAS (Mysore), officers, JCOs and other ORs who have supported and helped me in getting all the old records of Ajit and giving me the much-needed moral boost.

All the Mysore Marauders 18 MADRAS (Mysore) who have worked with Ajit, starting from his first CO, Late Col Om Prakash Chaman (Retd), I was fortunate to have an interaction with him about Ajit, as a young Lieutenant Ajit's DS in IMA, and later on his Unit officer, Maj Gen Dinesh Purshottam Merchant, AVSM (Retd) for constantly enlightening me about the Unit, narrating patiently about his tenures and Mrs Savitha Merchant for sharing her valuable observations about Ajit, Lt Gen Arun Kumar Chopra, PVSM, AVSM (Retd), Ajit's CO in Gandhinagar and Keri, former Col of the Regiment for always supporting me in my endeavour. My humble gratitude to Brig Swapan Kumar Sarkar (Retd), Brig G. Athmanathan (Retd), Brig OPS Pathania,VSM (Retd), Col Ravi Shanmugam (Retd), Col Diljit Singh, YSM (Retd), Col Benny Sebastian (Retd), Col Ishwinder Singh Sehdeva, VSM (Retd), Capt David Selvaraj (Retd), Lt Col Ritchie Ashwin S for sparing their time and sharing the professional details of Ajit, others like Sub Major S Binu, who was Ajit's buddy during the operations and had to take a couple of bullets while trying to protect Ajit. Col C B Binu, SC, for the valuable communication with the media and liaising with ADGPI. I am grateful to Nk B Sivasankaran Nair (Retd), Nk Parameswarn (Retd), Hav BR Manoharan (Retd) and Hav BN Kannan (Retd) for reminiscing the time they had with Ajit not only during the various military exercises but also on the field while playing sports together.

The Commandant of the Madras Regimental Centre, Brig Rajeshwar Singh, SC, SM and his team for facilitating and assisting in the shooting of the promo at the centre and the constant support they have extended to me at all times.

Maj Gen S V Thapliyal (Retd) who was the Commander, 63 Mountain Brigade, Nathu La Pass when Ajit was Brigade Major, for discussing his experiences with Ajit, which gave me a peek into the soldiering life of my dear husband.

The Officers of the 25 Rastriya Rifles - Brig Anil Gupta (Retd), Col Ramachandran, VSM (Retd) who was the commanding officer, when Ajit was second in command, Col B S Poswal, Col Shorav Nandy (then young Captains), JCOs and ORs during Ajit's time who not only shared their precious pictures of Ajit but also recollected various operations that took place in the unit. Hav Shivkumar (Retd), who was Ajit's helper in office and on the day of the operation, was asked to join the OP Faizalabad and was a witness to the daredevil act of Ajit. Thanks to the present CO, Col Kamakhya for being very prompt in his responses to my queries related to the battalion.

My heartfelt gratitude to the serving/retired officers whom Ajit had mentored during his tenure in OTA, Chennai as a DS to the 47th, 48th, 49th and 50th batch. Col Sada Peter (Retd), Col Sachin Tyagi, Col Vikas Dimri, Col PK Daka, Col Sachin Tyagi, Col Sanjay M Pradhan (Retd), Capt APJS Gill (Retd), Col Atul Pradyot (Retd), Mr Surinder Kumar (Boston, US) and others for narrating various incidents with their DS of Kohima Company during their initial years of training as GCs.

Brig P K Vij (Retd) and the team from Vij Publishing Group for all their persistence and working on numerous drafts to make this book what it is.

Ms Maya Chandra and Team Maya films in the making of the promo for the book. Maya had agreed to take up the project the very second, I told her about it.

Ms Radhika Shenoy, Editor, Ajit's niece for assisting me with the editing, and nuances of writing.

Last but not the least, the family, Col Pradeep Bhat (Retd), Capt Sudhir Prabhu (Retd), his wife Ms Shiela Prabhu, Ajit's brothers: Col Arun Bhandarkar (Retd), Mr Anil Bhandarkar, my brothers: Mr Narendra M Kamath , Dr Gopalakrishna M Kamath, and my dearest boys Nirbhay Bhandarkar and Akshay Bhandarkar for constantly sharing their critical comments and giving their honest opinions about the book, which helped me to refine it from the millennial's perspective and other family members for all the moral support and encouragement given to me in the making of the promo and the book.

There are many more people whom I would have loved to mention nevertheless my utmost and heartfelt gratitude to each and every one of them.

Glossary

18MM	:	18 MADRAS (Mysore)
ADGPI	:	Additional Directorate General of Public Information
Adjt	:	Adjutant
Battalion	:	A large unit of soldiers that forms part of a larger unit in the army
Battie	:	Helper
BM	:	Brigade Major
Sentries	:	Security Guards
CET	:	Common Entrance Test
CO	:	Commanding Officer
COAS	:	Chief of Army Staff
Coy	:	Company
DS	:	Directing Staff
GC	:	Gentlemen Cadet
Hav	:	Havaldar
HQ	:	Army Headquarters
IMA	:	Indian Military Academy
IN	:	Indian Navy
IPKF	:	Indian Peace Keeping Force
JC	:	Junior Command
JCO	:	Junior Commissioned Officer

MES	:	Military Engineering Service
MESS	:	A place where officers formally dine in.
MMG	:	Medium Machine Gun
MS	:	Military Secretary Branch of Army Headquarters
NDA	:	National Defence Academy
OR	:	Other Ranks
Paltan	:	Battalion
RR	:	Rashtriya Rifles
RCL	:	Recoilless gun
Regiment	:	A Military unit
Runner	:	Helper
SC	:	Senior Command
Sahayak	:	Buddy/Orderly during operations or otherwise
Sep	:	Sepoy
SFA	:	Separated Family Accommodation
Sqn	:	Squadron
SSB	:	Service Selection Board
SSBJ	:	Sainik School Bijapur
STF	:	Special Training forces
Subaltern	:	Senior officer from the same unit
UN	:	United Nations
Unit	:	Battalion
Walkman	:	Portable device to listen to music

Chapter 1

Introduction: The Birth of a Braveheart

In the Southern state of Karnataka lies a small town called Udupi, which is famous for the Krishna temple. Udupi Vasudev Bhandarkar, having born and brought up in Udupi, migrated to Mumbai, to serve in New India Assurance Company. It is here, as a young bachelor, that he got married to Shakuntala Acharya.

Udupi Vasudev Bhandarkar and Shakuntala Bhandarkar (both my mother-in-law and I share the same name) stayed in Wadala, Bombay in the early '50s. They were blessed with three sons, the eldest being Arun (Col Arun Bhandarkar, Retd) born on 5th April

Mr Udupi Vasudev Bhandarkar and Mrs Shakuntala Bhandarkar

Ajit with his mother, his brothers Anil to the left and
Arun to the right

1955, Anil born on 6th February 1958 and the youngest was Ajit, born on 31st December 1960.

Ajit, being the youngest among them, was very naughty and a super active child. He would feel secure in the presence of his mother while his mother was busy with the household chores. Though it was a nuclear family, their home was always packed with relatives coming over while on their way to Kasi and Haridwar.

The new economic reforms of the '60s gave a big boost to the Insurance Sector. Ajit's father was asked to go to Bangalore to open one of their new branches. So, in 1964 he landed back in Karnataka to start the new office of New India Assurance Company. The family stayed in the company accommodation at Rajajinagar. All the three Bhandarkar boys joined National English School; the present National Public School Rajajinagar, Bangalore.

Since Ajit showed interest in joining the Army, he was then mentored and counselled by Mrs Kusum Bhat, the Headmistress, and was successful in clearing the entrance exam to join the Sainik School Bijapur.

A family pic with 2nd Lt Arun Bhandarkar, Anil and Ajit (left)

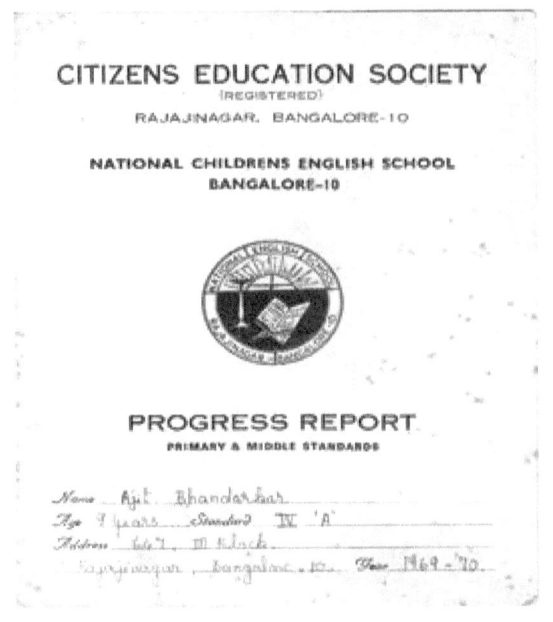

Ajit's report card, Grade 4.

Just like Jijabai had a huge influence on Chhatrapati Shivaji Maharaj for his bravery and courage, Ajit's mother, too was very encouraging, cheered him to do his best and motivated him to follow his dream. So much so that both her eldest son, Arun joined the Corps of EME and her youngest son, Ajit joined the Infantry.

Baby Ajit with his mother

Baby Ajit with his father

CHAPTER 2

AJEET HAI ABHEET HAI - SCHOOLING AT SAINIK SCHOOL BIJAPUR

Ajit was the youngest of the three brothers and wanted to join the forces since a very young age. Having been brought up in Bangalore, he did his primary education at National Children's School.

Ajit Bhandarkar standing second row, first from left with his cousin Sudhir Prabhu, sitting on the ground third from left.

This is presently the National Public School, Rajaji Nagar, Bangalore. He was very active and enthusiastic as a child and always shared his dream of becoming an army officer. Though we didn't have a single member in the family in the Defence Services, he wished to join the Army. Ajit's uncle, Mr Sadanand Bhat, the Chemistry teacher and Mrs Kusum Bhat, headmistress, Primary School, his aunt were serving in Sainik School, Bijapur (SSBJ). They motivated and guided Ajit to join the SSBJ. His mother, with a heavy heart, agreed to send him to the SSBJ. During his initial days, Ajit would miss his home, however with time he started enjoying his schooling. The students had roll numbers and to this date are remembered by their roll numbers. Ajit's roll number was 567 and belonged to the Hoysala house and later on became the School Captain, in 1977.

Hockey 1976   Discus Throw 1976

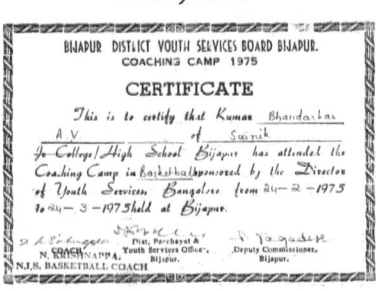 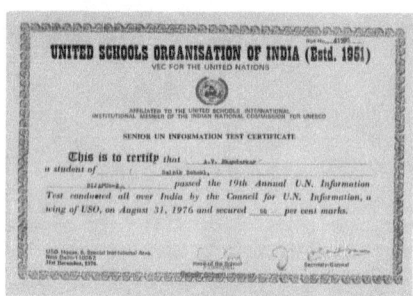

Basket Ball 1975   Senior UN 1976

Ajit belonged to a very close-knit family and the presence of his uncle, aunt and cousins would keep Ajit motivated and happy. As a student, he was very determined and focused and would want to give his best in whatever he did. His cousins, Pradeep Bhat, Deepak Bhat and Nivedita Bhat, were always there for him and a beautiful bond was formed, which continued even after the school days.

(L to R) Pradeep, Nivedita, Deepak and   *Mrs Kusuma Bhat, second from right*
Ajit at the Staff Quarters   *with Ajit standing right at the back*

Here, as an educator, I would like to emphasize on the importance of schooling in the life of a person. Life in SSBJ was regimental with the early hours of PT and drill, followed by breakfast. They would then attend to their classes till evening with a lunch break in between. The evenings were spent playing team sports. Ajit, besides his academic excellence, was the school cricket team captain. He played full back in football and hockey, and it was difficult to break his defence since he was so proficient in his skills. Apart from that, he also loved horse-riding, swimming and volley ball. Since he was also the School Captain, his team spirit and leadership qualities were gradually getting sharpened during his adolescence. He was so genuine and empathetic that his classmates called him 'Manav' (idealism) - in memory of the idealistic character role that cine star Dharmendra had played in the famous Hindi film *Dost* in 1974.

Mr Gopal Kulkarni, (Roll no. 546), Classmate of Ajit recounts: "This was in 1977, the final year of our school. Ajit was selected as the School Captain - the highest achievement any student could achieve. The role had its own challenges. Maintaining the discipline and ensuring that all activities take place smoothly was a big responsibility, given that there were more than 500 students of varying ages. Often, he had to be firm and tough on the students. But when it came to his own classmates, he was careful to be respectful and yet get things done."

"One afternoon I bunked the preparatory classes and slept through. This would normally result in a punishment. But upon finding that I was missing, Ajit walked up to me and asked what was the reason. I told him, I just slept through and did not wake up in time. He told me not to repeat this, and to ask someone to wake me up the next time. Knowing that I was his classmate, he handled me differently yet effectively. Touched by his respectful behaviour, I never again bunked the afternoon preparatory class. This showed his humility and the fact that he had become school captain did not make him high headed. He continued to maintain good friendly relations with all his classmates. With this, he truly showed his mettle."

Like many of his classmates, he too cleared the UPSC exam, in January, 78 and joined the 59th Course of NDA.

Sl. No	Course No	ROLL No	Name Of the Cadet	Month & Year of Joining	Wing/Branch
94	55th	346	SR BHARAMANIKAR	Jan.1976	ARMY
95	"	350	SB KARADI	"	ARMY
96	"	369	RS YADAV	"	ARMY
97	"	455	SG BALUTAGI	"	ARMY
98	"	462	PD HALLUR (OTS)	"	ARMY
99	"	464	SR MARANABASARI	"	ARMY
100	56th	330	JC KAMBALIMATH (OTS)	"	ARMY
101	"	377	IS SUHAG	"	ARMY
102	"	381	AK RAGHAV	"	
103	"	403	MB RAVINDRANATH	"	ARMY
104	"	414	RR KULKARNI		ARMY
105	"	420	BY KONNUR	"	ARMY
106	"	485	SA PANDE (OTS)	"	ARMY
107	57th	337	RB PATTANASHETTI	Jan.1977	
108	"	371	RAJPALSINGH AHLUWALIA (OTS)		NAVY
109	"	379	ASHOK KUMAR (OTS)	"	NAVY
110	"	436	SA DHANASHETTI	"	NAVY
111	"	586	HT JAGADISH	"	ARMY
112	58th	402	SB SAJJAN	Jul.1977	ARMY
113	"	439	VK CHHATRE	"	ARMY
114	"	475	SS DHAKE	"	IAF
115	"	500	KP BHAT	"	ARMY
116	59th	512	S NEERAJ ROY	Jan.1978	ARMY
117	"	517	SRIKANT	"	NAVY
118	"	535	CS MULGUND	"	ARMY
119	"	567	AJIT BHANDARKAR	"	ARMY
120	"	583	RAJESH MALHOTRA	"	ARMY
121	60th	533	ANNIRUDHA N GUDI	Jul.1978	ARMY
122	"	543	BALKRISHNA DILIP	"	ARMY
123	60th	598	SB VAJJARAMATTI	Jul.1978	ARMY
124	"	608	SV DESHPANDE	"	ARMY

The Roll of Honour Board at Bijapur Sainik School

(L to R) Col Murali, Principal SSBJ, Ajit's mother, and I during our visit to the school with the aspiring students

The school has literally groomed and contributed in the making of a host of Generals, Air Marshals and Admirals: Lt Gen Ramesh Halagali PVSM, AVSM, SM (Retd), Late Air Marshall Sriram Sundaram (Roll no. 2), Maj Gen KN Mirji, VSM (Retd), AVM Yajurvedi, AVSM, VM (Retd), Lt Gen PG Kamath, PVSM, AVSM, YSM, SM (Retd), Maj Gen KS Kumbar, VSM (Retd) (Roll no. 158), Maj Gen Arjun Muthanna (Roll no. 220), Maj Gen VS Gaudar, Rear Adm Purandhare, Late Vice Admiral Srikanth, AVSM (Roll no. 517), IGP Gopal B Hosur, Retd IPS, (Roll no. 177) and others.

The Batch of '77 is so fond of Ajit that they all joined together and dedicated a Dwar, in his name, on 1st Jan 2019. The idea of Lt Col Ajit Bhandarkar Dwar was conceived during the batch meeting with the Principal on 15 Sep 2017, when they donated an amount of Rs 5 lakh. When Col Biswas, the then Principal, solicited suggestions for utilizing their contribution, Col Anirudh Gudi (Retd) (Roll no. 533) had promptly suggested the construction of Ajit Dwar to commemorate his martyrdom and gallantry. Immediately a committee was formed and a team of his classmates took on different roles for the construction of the Dwar.

The finances were met by Ajit's batchmates, who contributed towards the project of which - Col K Pradeep Bhat (Retd) (Roll no. 500) was the Treasurer and the architectural design was given by Mr Jagdish Nandi (Roll no. 566). Execution of the project was done by GS Patil, (Roll no. 521) Executive Engineer, while the local coordination and supervision was handled by Mr Arjun Misale (Roll no:520) and Mr GS Patil. The guidance and support were given by Late Vice Adm M Srikant (Roll no. 517). Both the ex-Principal, Col T Biswas, and the current Principal, Capt V Tiwari, were instrumental in the implementation and completion of the project.

Local Media Report on the Ajit Dwar

'77 Boys Group Photo - 2019

So, the bonding of the school-days was so strong that their friendship is evergreen. Every time I meet them, they always have some incident or the other to share with me and fondly remember their '*manav*'.

RIVALRY WITH A GREAT SOULMATE FRIEND

Ajit, popularly called *Manav*, and I were a one-sided, on my part only, rivals at school, till the tenth std, since I was competing against him to be the School Captain, at Sainik School, Bijapur. We competed all along, being in the neighbouring houses, Hoysala and Vijayanagar, on the sports field, in academics and co-curricular activities, such as declamation contests, drill competition and so on.

The surprising thing was, in the eleventh standard Ajit and I, went on to become the best of friends. The transformation in me was a miracle and Ajit's acceptance of a rival friend, a friend. It was certainly a miracle to which I owe my gratitude to Pradeep Bhat (Col Pradeep Bhat) and Ajit (both cousins).

Our last day in school and the train travel from Bijapur railway station to Hubli is etched in my mind. Ajit's help in guiding me cannot be forgotten. I distinctly remember my stay with the Bhandarkar family in Rajajinagar, when I had gone to take our second NDA exam attempt. How can I forget the hospitality of Ajit's mother and the quiet calmness of his father? The company of his brothers Arun and Anil is still fresh in my mind. I have very fond memories of my stay at his house.

It was predestined that we be together. Sure enough, we were in the same squadron in our first term in NDA, Ghorpuri, encouraging each other to participate and compete with the rest of the course mates from other school backgrounds. He excelled in hockey whilst I performed in cross country and athletic events. We spent our free time together.

Following the first term break, we were again destined to be together. Couldn't have asked for anything better, we were again together in Golf Sqn for the rest of the five terms threaded together with Pradeep Bhat who had joined the 58th Course, our senior in the same sqn. We spent the best time of our lives with the 59 Golfies. We were trained very hard with the Golf sqn traditions. In the winter of 1980, we parted ways, Ajit moving to Dehradun to IMA, whilst I proceeded to EFS, Bidar since I was an Air force cadet. I was not destined to be a flier and hence requested for a change over to the Army.

Back into the Army, I was fortunate again to be with the 59 Golfies, though I lost a year, in July of 1981. I had, by now, started taking life as it came along, one day at a time. Ajit and I met off and on at IMA, on the weekends. The 59 Golfies passed out from IMA, in the winter of 1981. I stayed back to be commissioned in December 1982.

The Almighty once again gave us a chance to serve together in the same Division. Ajit was in Gandhinagar and I was in Udaipur. We met frequently in the deserts since we were in the neighbouring formations. In the Ahmedabad riots of 1985, we were once again deployed together, sharing our experiences.

There was a void in our meetings from 1986 until 1988 when I moved to Sri Lanka, as part of the IPKF. I remember visiting Ajit in OTA, Chennai, where he was posted as a DS. We spent an evening together. I distinctly remember him seeing me off at Chennai railway station when I was proceeding to Bangalore.

We were in touch through letters. Sometime, in 1993/94, Arjun Subramaniam and Ajit visited us at Pune, when they were on the DSSC tour. Kumuda, my wife, met Ajit for the first time. She was always anxious to meet my good old friends and classmates.

Between 1995 to 1999, I had the opportunity of meeting Ajit many a time. 1995 Ajit's birthday celebrations, I can't forget with Brig Pattar, Ajit and I at the Army Battle Honours Mess, Delhi followed by Chinese dinner at Malcha marg and then the New Year celebrations at the India Gate. In 1998, I too got posted to Delhi. I was the one who bought Lt Col badges of rank from Gopinath Bazar for Ajit, when he was proceeding on promotion to 25 RR. Rest is known to the world.

My take-away from Ajit - Stay cool, calm and composed. Be an Officer and Gentleman, blessed to be his friend, can go on and on.

Lest I forget, his autographed words to me - "We are a speck of dust on this vast dreary desert of sand."

His Intelligence collated diary which I happened to get in Bhaderwah, helped me to eliminate many more terrorists in that area!

Lastly, indeed grateful to the Almighty for having bestowed and honoured great friends like Ajit. With gratitude and salutations to a very noble and peaceful soul. I look forward to meeting Ajit, wherever, I can find him.

Om Shanti, Ajit.

Col Som Neeraj Roy (Retd)
Corps of Intelliegence

AJIT - AS A CLASSMATE

26 June 1970 is as unforgettable to me as it is my birthday, and also later, my wedding anniversary day. It's the day, when we both, Ajit and I, joined Sainik School Bijapur (now Vijayapur) in the fifth standard, leaving behind our loving parents and adoring siblings at the age of nine. Ajit was placed in 'A' section as he had studied in English medium and I was placed in 'C' section as I had studied in Kannada medium till the fourth standard. In the first year, our off-class interaction was also quite less as he was in Hoysala House and I was in Adilshahi House. My earliest recollection of Ajit is of a sincere, diligent, sober and quiet boy who was somewhat aloof.

We became closer from our sixth standard onwards as I was shifted to 'A' section but our interactions were more inside the class than outside. We liked each other but had our differences too. I chose Sanskrit as my second language while Ajit chose Kannada till our eighth standard and thereafter, I chose English Literature while Ajit chose History as an additional subject besides the compulsory subjects Physics, Chemistry, Mathematics and English. As we came to senior classes in our teens, Ajit turned out to be an all-rounder with good grades and proficiency in sports. Ajit also grew tall and built a good physique but never showed off his strength or bullied anyone.

My lasting impressions of Ajit are of his insistence on propriety in all actions. He never let temptations of childhood or teenage affect his ideals and values. This aspect of his personality was so evident that he was identified by all with the role of an idealist called 'Manav', played by Dharmendra in Dost, a Hindi movie released in 1974. 'Manav', which became Ajit's nickname thereafter, was a portender of things to come in school as well as in his career leading to his ultimate sacrifice. Ajit was appointed the School Captain for his idealism, values, all-round performance and leadership qualities. He handled his duties with poise and dignity even in the face of some extreme provocations, thereby winning the support of one and all.

In the National Defence Academy (NDA), we were in neighbouring squadrons though in different courses - Ajit in Golf

14

59th course and I in Foxtrot 60th course. Our mutual respect continued to grow though we parted in our different paths. Ajit went on to be commissioned into 18 MADRAS (Mysore) in December 1981 and I joined the Corps of Signals in June 1982 from the Indian Military Academy (IMA). Thereafter, my wife, Vinodini, and I had the opportunity to attend some of his post-wedding rituals at Chitrapur Math in Bengaluru in 1990 and meet his charming wife Shakunthala. Fast forward to 1997, when I was posted in Sri Ganganagar after completing my staff course at Wellington and Ajit was posted to the Military Secretary's Branch in Army Headquarters after completing his tenure as a Brigade Major. Ajit picked me up along with my parents, Vinodini and kids - Amogh and Apoorva - from Nizamuddin station in his Maruti Omni and hosted us for half a day at his house. We enjoyed Ajit's hospitality and the sumptuous lunch prepared by Shakunthala and played with little Nirbhay and Akshay. Alas, it was our last meeting and remains etched in the memories of my wife Vinodini, my mother and me.

Though I admire Ajit's ideals and sacrifice, I also keep wondering whether his life was worth the lives of a few worthless terrorists. I know Ajit won't agree with me, as he is *Manav*.

Col Aniruddha Narashimha Gudi (Retd)

Corps of Signals

THE AJIT I KNOW

Ajit V Bhandarkar, known him since 1970, when he joined the school in fifth standard and I was in the sixth. It is just over five decades now...half a century! We were in the same house in school and in the same squadron in NDA and I had been in touch with him during our Service and it was in Army HQ that we were posted together (1997 -1999), he at the reputed MS Branch and I at MO Directorate. It is here we had a very close interaction both professionally and socially. Our children were of the same age and we would very frequently meet and that is when I really got to know and understand Ajit.

It is extremely unfortunate that he is not amongst us today, if at all if anyone deserved to be here on this planet... it is Ajit. He was an epitome of life. From the days I have known him, he has been very humble, sincere, extremely dedicated and good at everything including sports and academics. A very soft-spoken, affable and down to earth person. The biggest problem I had as House Captain with Ajit was, he would never argue and speak loudly, but solve or suggest solutions in his own simple and practical way. He had such a clear understanding of things that, he was always right and to the point. Somehow, we grew to be fond of each other...and continued to share a very affectionate bond right through. Even to this day...I miss him.

Ajit, actually was like a hard, solid stone. He would be there in every activity of the House in school and always actively participated in almost all of them. His very presence meant that things would be fine as he used to make sure that not only he gave his 100 per cent but also ensured that others too followed his example. There was something about Ajit that made him stand apart in his school- days...difficult to say exactly... probably his simple and charismatic personality or his humble approach or his sheer simplicity and willingness to share and participate in every activity or always approachable or just his tall personality! He just excelled in everything and was still so humble that it was difficult for many to understand how he never spoke or boasted of his achievements to anyone. He just went about doing things in his own simple and dedicated way. One thing for sure he was respected by the seniors in school, classmates adored him and his juniors did maintain their distance and looked up to him as an ideal senior.

Even at NDA, Ajit continued to participate and excel in all activities. We continued to interact on many issues and in those four-odd semesters that I really saw him grow and mature. At NDA, we were in the same Golf squadron and he was just a course junior to me. We together participated in many activities; we were together in many 'ragada' (punishment) parades by seniors. In the final semester, we did go on out-pass to Pune city. Right through the tough academy days, Ajit was the same simple, solid, studious,

16

dedicated and focused boy. He was always full of energy and always had simple solutions to many complex issues of life. Whenever we discussed personal issues, he used to have a very simple logic and simple solution. Those were our teens and we did have dreams, fantasies and opinions on everything on earth but Ajit always toned me down to earth. Maybe this was why we were very fond of each other.

After NDA, next we met just a few days after my marriage at Bangalore. I was on a short leave and so was he. I was busy doing my shopping. He dropped by one day and heard my shopping list and places that I was buying from. He got furious and annoyed. For the next four / five days he accompanied us and ensured that we bought what was just required and from the right place and at a reasonable price. Thanks to him I got wise on spending, on day three I bought a wall clock costing Rs. 300 or so, when he heard that he blasted me and made me return it. Without informing me he went to one of his friend's place and got me a clock. All this while he was hard up for time for himself. On the last day of our leave, I still remember the farewell dinner at Nagarjuna on Church Street. We enjoyed the lovely Andhra meals and also Ajit's effort to convince me on saving money, managing home and family, etc. I was really surprised by his understanding, maturity, his genuine and sincere concern for me.

It was in Delhi that we got to meet very frequently. Professionally, he was a topper and was posted at MS Branch and I was in MO Directorate. We were in our mid-career stage and in our middle age with two young children. Though posted in MS Branch, he never had any airs about it. He treated it like any other normal posting and went about it in his own dedicated and serious way. This was highly appreciated and for this, he was held in very high esteem by all of us. Many a time he used to tell me to slow down and take life normally and not be too concerned about promotions, etc. That time I was in a high-pressure job and meeting Ajit was always a comforting experience. He would always and every time convince me to just work honestly without expecting anything in return. I clearly remember he told me that life will take its own course and not to bother too much about it. Just prepare yourself and the family to face life. Those were some of the golden advice I got and have followed it sincerely and it has come handy in dealing with my own personal tragedy. Many times, I wondered, how Ajit was so simple, mature with so much understanding of life.

Socially, my family and I used to visit Ajit and family very often for dinner on weekends as we had brats of the same age. A few times just the two of us went to Lodhi garden on Sunday morning with our children and spent that morning chatting. My last and fondest memory of Ajit was when I had invited their family along with other Ajeets in Delhi for dinner. It was August '99, we had a blast of a time. Everyone roaring with laughter and it was good fun.

The shocking news I got while I was in Namibia, Africa on a UN peacekeeping mission, was unbelievable. I truly lost a great friend and miss him very dearly. It is my personal loss too.

Om Shanti

Brig SB Sajjan (Retd)

Mahar Regiment

CHAPTER 3

SEVA PARMO DHARMA: NDA DAYS

Way back in 1917, the Secretary of state of the British Government, in a statement in Parliament, had conceded to Indianise the army and establish an Academy for training Indian officers. But no implementation was in sight. The British Indian Army continued to be entirely British in the officer cadre and a mix of British and Indian in rank and file. It was only in 1922, the Indian Officers, were sent to Royal Military Academy, and were called the King's Commissioned Indian officer (KCIO). The India Military Academy was established in 1932 as an essential concomitant to the policy of Indianising the army.

So, the idea of a 'Joint Services Academy' was conceived even before our Independence; i.e., in 1945, with the end of World War II. Three things had become very clear: Indian Armed forces had to expand in a big way; an institution for training these officers had to be designed and established; and last but not the least, the British Indian Government will cease to exist. So, therefore, a 'Nation War Academy' (the suggested name) working committee was formed.

With the attainment of Independence, on 15th August 1947, national security became a matter of paramount importance. Thus, was established the National Defence Academy (NDA) of India. Since the project for the construction of campus would have taken a long time, as an interim arrangement, training started at Clement Town, Dehradun, where the Inter-Service wing (NDA) functioned. The training session was inaugurated by Sardar Vallabhai Patel, on 4th June 1949 and the foundation stone for the world's finest premier institution for the Military inter-services, was laid by Pandit Jawaharlal Nehru, at Khadakwasla, Pune, on 6th October 1949.

The present NDA was then formally inaugurated on 16th January 1955 by Shri Morarji Desai, the then Chief Minister of Bombay state (which included present-day Maharashtra and Gujarat). This great institution, the National Defence Academy, was, therefore, a landmark in the history of India's defence forces and also the country's post freedom achievements. In today's terminology, the Atma Nirbhar Bharat!

Khadakwasla was chosen because of its proximity both to the sea, and also an established airbase. Apart from these factors, it also used to be the hunting grounds of Maharaja Chhatrapathi Shivaji, that would provide an excellent location for all the military exercises by cadets.

Almost every student in Sainik School endeavours to join the NDA. Having cleared the UPSC exam after his Grade 12, in 1978, Ajit joined the 59th course of NDA as part of the Golf squadron. Many of his classmates too joined the NDA which further strengthened their bonding and camaraderie.

Here again, at NDA there was a rigorous training routine, which started with physical training and drills followed by classes and then other sports and games in the evening. All juniors were at the receiving end of their seniors' *ragda* (punishment) during the initial few terms. Any human being's nature would be to loathe such 'ragda sessions' but these were instrumental in building the esprit de corps amongst the peers while also forging long lasting bonds with seniors. Front rolls, back rolls, knuckle push ups, hiking with heavy rucksacks to the top of Singarh Fort were common punishments for the cadets. The cadets sometimes were told to climb the hill, carrying their cycles in their hands, above their heads if they didn't follow the rules of the Academy, which was one of the ways of disciplining and toughening them.

While the physical training, which included drill, swimming, parade, horse-riding, team sports, individual games and cross-country runs, built up the stamina and strength of the cadets, the academic curriculum of university standard embraced a comprehensive syllabi, in foreign languages, English, Maths, Science, Social studies, Engineering Drawing and also purely Military subjects like Military History, Military Geography, Tactics, Field Crafts, shooting, weaponry, etc. enlightened the young officers in the making.

Apart from their regular roles and responsibilities, cadets were selected for junior appointments and taught the tenets of leadership deliberately from their third term onwards. Basic leadership was assessed not only by watching how a boy could handle a group and win their confidence in their day-to-day activities but also giving them important roles during their camps and training exercises. Ajit, in the sixth term, became the Sergeant of his Golf Squadron and also participated in team sports like hockey, football and basketball and individual sports like boxing.

Ajit (3ʳᵈ from left sitting) along with his cousemates

Ajit (2ⁿᵈ from left) standing with coursemates from Golf squadron and his div. officer

21

After completing the six terms, the cadets are mentored and trained to join the Indian Military Academy, Dehradun. The much awaited passing out parade takes place after each course where in parents are also invited for the event and the most accomplished cadet is presented with the Sword of Honour, during the ceremonial parade followed by marching past the *Antim Pag*.

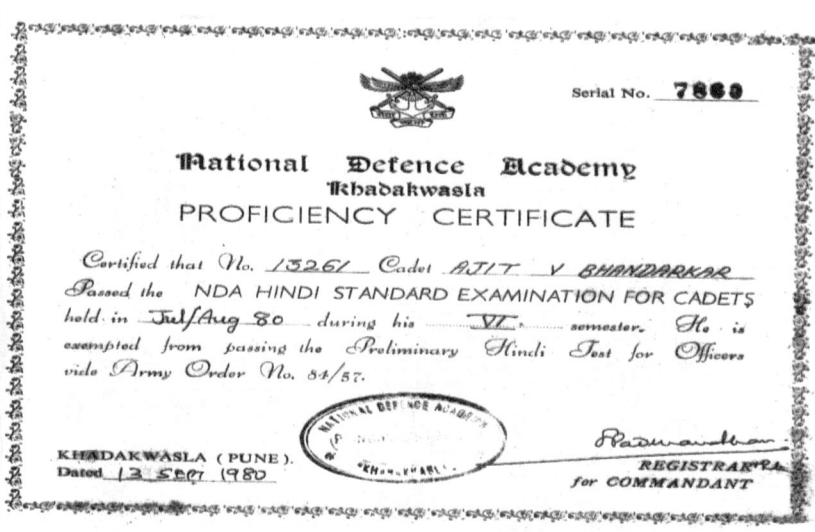

Hindi Certificate - 1980

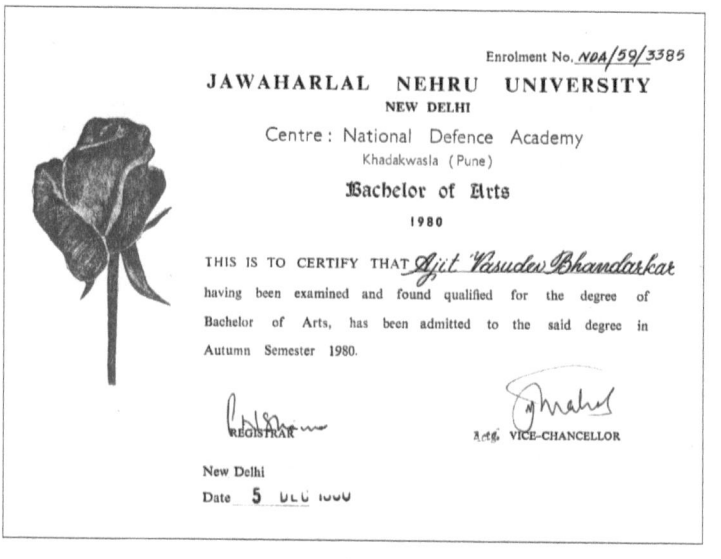

BA Certificate - 1980

One of Ajit's cousin, Sudhir Prabhu, who was his classmate in National English School later joined him in NDA and IMA. They both later were commissioned in the Madras Regiment, while Ajit joined 18 Battalion the Madras Regiment (Mysore), Sudhir joined the 3 Battalion, The Madras Regiment.

Here is what Capt. Sudhir Prabhu (Retd) had to say about his dearest cousin Ajit:

"Between the two of us, I was the temperamental one, expressing my likes and desires, wearing my aspirations on my sleeve, whilst he was the calm and composed one. His focus, *josh* and sincerity in that order which earned him respect amongst the course mates, as one of the most level-headed buddies."

Till date Ajit's course mates not only keep him in their thoughts but also continue to constantly stay in touch with me and offer their help and support whenever required. Many beautiful memories of the academy days have been shared by some of Ajit's Course mates.

"His decency did not allow him to take initiatives. Had the josh of a horse." — *The NDA Journal Vol: 26*, Autumn term 1980

Ajit – You Were Selfless and Uncomplicated

It was in January 1978 that Ajit and I joined the National Defence Academy at Ghorpadi for our training of the first semester before moving out to Khadakwasla for the balance training. Like us, there were another 300 who had joined from different states, diverse schooling background and other contrasting features. Ajit and I were allotted the same division in November Squadron at NDA wing Ghorpadi and were housed in barracks each accommodating about 8 to 10 persons. Ajit and I were in different barracks; however, the nature of training and curriculum being group-oriented, our interaction was very frequent. The curriculum was challenging and physically very demanding. It is natural for the 300-odd teens immediately after their class XI to start searching for like-minded people to share some quality time together to mitigate the harshness of the training regimen – birds of the same feather flock together.

Ajit and I were both quiet and certainly not much enamoured by the flamboyance of any kind manifesting in any form. Though we never knew then, but as I think of it now, possibly we valued situations similarly and therefore choose to spend time in the company of each other when not committed to training activities.

Ajit had profound qualities of mind and heart. Generally, a person tends to either like or hate or be indifferent to a person. I did not come across any person who disliked or was indifferent to Ajit. He had very high self-esteem and enjoyed the trust of all. True as a leader, he was friendly but avoided familiarity. He always meant what he spoke, rarely judgemental, and always gave honest opinions. That may be the reason for his quietness – still waters are deep. He was a man of substance as events in future would substantiate. Today as I raise my pen to write after four decades, memories of his sincerity, steadfastness, warm-heartedness and trustworthiness overwhelm me.

As luck would have it, both of us again found each other together in Golf sqn after moving to NDA Khadakwasla from NDA Wing and our friendship further strengthened. Being in early teens and from a boarding background, my worldly awareness was rather poor and it tended to be either black or white. The grey area in between did not exist in my scheme of things. I found Ajit's maturity much beyond his years and it was rather rare that I ever found him reacting. However, when deemed appropriate, he did respond. It was generally believed that cribbing at NDA was everyone's birthright and it was promoted possibly because it mitigated the harshness of the curriculum. So long as it remains healthy and fun it is fine but many a time it tended to be carried too far and those were the times when affectionate friends like Ajit helped you reflect on reality and the correct perspective. Ajit was a man of great character. We were in our fifth term and during that term the Inter squadron Hockey Championship was scheduled. Each sqn had to field six strings. Participation in games and sports were graded based on which string you represented the sqn. Hence a person in the second string was graded better than third string. Ajit was a much better hockey player than me and we both knew it. During the evening fall in the lobby of the squadron ground floor, the strings for the championship were announced by the Squadron Hockey Captain. Ajit's name figured in the third-string while mine figured in the second string. Since both of us were standing together, we looked at each other with disbelief. Notwithstanding, he gave a befitting display of his skills during the matches which ensured a very high finish of his string in the championship. There was no brooding or ever a mention of this thereafter. I admit that possibly my reaction would not have been the same as his – it would have tended more on the negative side. Team spirit and esprit de corps displayed by Ajit always and every time were profound. Rarely would Ajit be found missing from the late night course bathroom punishment sessions at NDA. During the various camps, he would always volunteer for additional errands which the majority preferred to shirk. I do not recollect Ajit ever doing individual punishments like restrictions or extra drill. It was just not in him to cut corners or resort to shortcuts. Ajit truly epitomised the NDA prayer, always choosing the harder right instead of the easier wrong. That was his way of life.

On completion of our training at NDA, we moved to the Indian Military Academy, Dehradun, as Army Cadets during January 1981. At IMA we were allotted different battalions. The training at IMA was organised Battalion-wise except for Central lecture / Demo and thus meetings got fewer, however, we would catch up whenever we got a chance to meet. Ajit did very well at IMA and passed out as a tabbed appointment, high in merit and his coy Cdr ensured that he got commissioned into his unit. The coy Cdr himself was a person of great repute and held in awe by Gentleman Cadets and I am sure he would never have regretted his decision. Ajit did him proud.

Post-commission, we did not have the fortune to serve together in the same station or formation, however, remained in touch through common friends and acquaintances. After all, those were not the days of Facebook and WhatsApp. After having met him at Belgaum while doing Young Officer Course during December '82, I met him again in November '96, at New Delhi. He was posted in MS Branch and was residing at Kaka Nagar. I was in Delhi to attend a family wedding and could not miss meeting him. I landed up at his residence as it was a holiday at around 1000hrs and spent an hour with him to catch up on our life events, prior to joining the wedding celebrations.

The next year, summer of '97, Ajit was visiting Dehradun with his family and we spent an evening at our place at IMA Dehradun. I was leaving for Wellington shortly to do the staff course. My staff course got delayed as it took time to get the waiver of not meeting the eligibility criteria. Ajit, that night, shared with me (since it was no more confidential) that he was aware that my case was under consideration and hoped like hell that the decision got ruled in my favour. He was very forthright in admitting that he had no role in the processing of the case. I am absolutely sanguine that he would not have influenced the outcome of the case in anyway but must have prayed for my wellbeing. That was my friend Ajit-rest in peace buddy. Shakunthala, Nirbhay and Akshay have continued his legacy forward with elan and dignity.

Brig Ved Prakash (Retd)

Principal Director and Secretary

Army Group Insurance Fund

GOLFIE AJIT: CLASSIC CASE OF 'STILL WATERS RUNNING DEEP'

Ajit Bhandarkar was a classic case of still waters running deep, a man with unlimited depth of character, an excellent human being and a comrade like none other.

I first met him in the barracks of Ghorpadi in January '77. He stood out as he was a disciplined soldier. Then destiny brought us together in Golf Squadron of NDA. In the early terms, he remained confined to his quiet world. It was only in the final term that we developed a mutual respect for each other in the wilderness of the Army Training Team, in olive green overalls with freshly painted packs on our backs.

His character qualities were illuminated during the Camp of the final term. He was a live wire, a comrade who willingly took upon himself to ferry water to the hilltop defences, who put in more than his share in digging trenches and who helped in preparing the briefings even if they did not involve him. The march back to NDA was an overnight exercise. I recall vividly, as dawn broke, we were still a few hours away from NDA. I was seated on a rock, taking a break from the march when Ajit appeared on the skyline. He was silhouetted against the sun, a rifle slung across one shoulder, a Light Machine Gun across the other. He had been carrying the LMG for the most part of the night without complaining or whining. I had to request him to part with it for the remaining part of the march.

The day we became Officers, Ajit asked me to put the lone star on one of his shoulders. That was the ultimate honour he bestowed upon me.

We met at Belgaum during the Young Officers' Course and it was Ajit next door who was most knowledgeable amongst all Golf mates. He had the most balanced head in the Golf squadron barrack.

My best interaction with him took place at Mhow. We were undergoing the same Mortar Course. He had come well prepared with coffee arrangements and we would often share coffee at midnight after we had finished studying for the day. When I was selected to conduct the Confirmatory Instructional Practice, Ajit prepared the charts so that I could practise the oratory.

Unknown to us, God had a different plan for Ajit but he went down with all guns blazing, like a true soldier, an epitome of bravery. I have not come across a better officer and a gentleman than our immortal Ajit.

Col Rajesh Tiwari (Retd)

JAK LI Regiment

CHAPTER 4

VEERTA AUR VIVEK – IMA: THE DECISION TO JOIN THE MADRAS REGIMENT

The safety, honour and welfare of your country come first, always and every time.
The honour, welfare and comfort of the men you command come next.
Your own ease, comfort and safety come last, always and every time.

— *Field Marshal Philip Chetwode*

The IMA credo is inscribed in the oak panelling at the eastern entrance of the Chetwode Hall is the Academy's credo, excerpted from the speech of Field Marshal Chetwode at the inauguration of the Academy in 1932.

Ajit joined Indian Military Academy, 69th Course, after graduating from National Defence Academy. There are other ways of entry to this prestigious institution, which are on graduation from Army Cadet College (a wing of IMA itself), via direct entry through the Combined Defence Services Examination followed by SSB exams, and as a technical entry under university and college schemes. While those who gain entry into IMA go on to become permanently commissioned officers, those who go to the other officer training academies such as Officer Training Academy, Chennai and Gaya, are conferred with the short service commission. Lady Cadets are not inducted into the Indian Army through IMA, though there are a few lady officers who are instructors in the Academy. IMA has a sanctioned capacity of 1,650 and the GCs get a stipend amount of Rs 56,100 per month.

The Gentleman Cadets (GC) of IMA are organised as a regiment with four training battalions, consisting of four companies each. Battalions are named after Generals of the Indian Army (except for Siachen Battalion), while companies are named after battles in which the Army had participated. So Ajit joined the Bhagat Battalion, Singarh company in January 1981.

Technical graduates, ex-NDA, ex-ACC (Army Cadet College) and university-entry cadets undergo training at IMA for one year. Direct-entry cadets train for one and a half years while the Territorial Army officers course is three months long. The core values of training are constant features in the entire spectrum of one's career, beginning with a cadet up to the rank of a General. These core values form the bedrock of the core objectives of the training of the GCs in the IMA and are echoed in the IMA Crest, Credo, Honour & Warrior Code and training motto "*Nischay Kar Apni Jeet Karoon*", which means determined to win. The core value of Competence is enshrined in the IMA Crest "Valour and Wisdom". The Honour Code of IMA "I shall not Lie, Steal or Cheat, Nor Tolerate those who do so", truly resonates the importance of the core value of character while the "Warrior Code" of IMA which has been adopted from the "Bhagwad Gita", the punch line of it is "I am a Warrior, fighting is my dharma;" also talks of compassion and valour.

Training is broadly categorised into character building, service subjects and academic subjects. Apart from the general subjects and military subjects, a sophisticated and state-of-art technology in weapon training is imparted to all the GCs, which includes Close Quarter Battle (CQB) range, Location of Miss and Hit Target System (LOMAH), Team Battle Shooting Range (TBSR) and finally Jungle Lane Shooting. Games, sports and PT, as usual, form the integral part of their activities throughout especially for an infantry officer. Ajit continued to play hockey, football and basketball in the academy too.

Ajit was fortunate enough to be mentored by one of the best officers, in the Army, Maj General Dinesh Purushotam Merchant, then Maj Merchant, who belonged to the 18 MADRAS (Mysore) unit of the Madras Regiment. Maj Merchant was the company commander of GC Ajit, and he had spoken to the GCs about the glory of the Madras Regiment and the calibre of Thambi troops. All this impressed GC Ajit and with a mentor like Maj Merchant, he was deemed and fortunate to join the oldest Regiment of the Indian Army. The Madras Regiment was raised between 1758-1766 to protect the Coromandel and Malabar coast. It fought numerous wars and its rich history, valiant officers and daredevil jawans inspired Ajit to join the Madras Regiment. Maj Merchant was also overwhelmed to see GC Ajit opt for 18 MADRAS (Mysore), which also had a vacancy in December 1981.

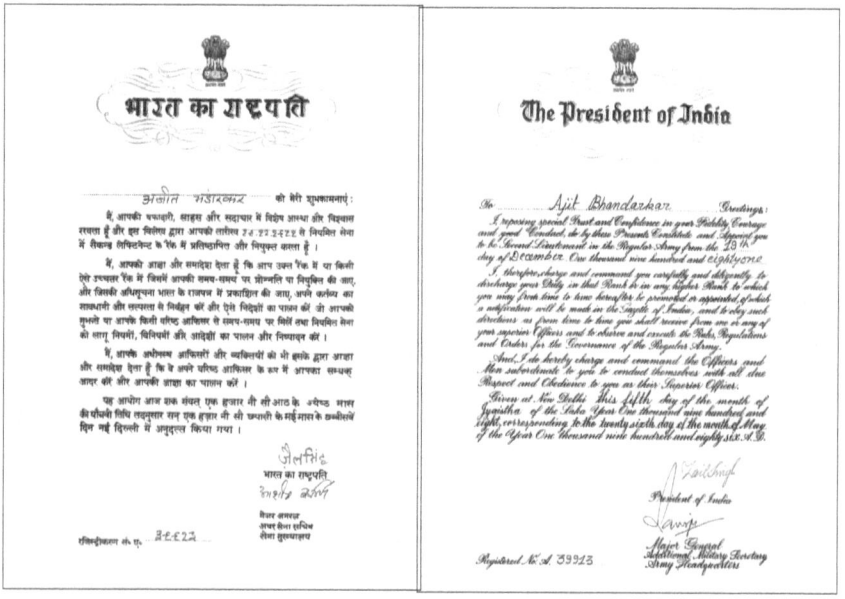

Formally commissioned into the Indian Army

Ajit's relationship with Maj Merchant grew as he joined his battalion, which is described in the coming chapters. Ajit not only had high regards for Maj Merchant but also his wife, Mrs Savitha Merchant. In IMA, the Mela is an annual feature, wherein all the company commanders are required to set up a stall and sell the goodies they made. She recalls the days when she was in charge of the stall where she was selling *Pav Bhaji* and *Kanda Poha*. While the cadets and Directing Staff (DS) were busy enjoying the goodies at the counter, she realised that she was running short of coriander which is an important ingredient for the garnish of the food items. When Ajit who was helping her as a cadet was told about it, he immediately rode on his bicycle to the market and got her the much-needed condiment for seasoning.

The training days, those events and the interaction with all these officers and their families slowly created a close bond and these relationships bloomed to form stronger ties progressively.

My Coursemate, Squadron Mate, Company Mate, YO Course Room Mate, Senior Command Course Flat Block Mate

My earliest memory of Ajit goes back to January 1978 when I joined the NDA 59th Course at Ghorpadi. I noticed him in the hockey field, playing a very fair game with a lot of skill. Realised later, on further interaction with him that fairness was ingrained in his blood, he did not have a single unfair cell in his body. My closer interaction with him began in the second term when I joined Golf Squadron and Ajit was also there. Now Squadron mates are a breed which is much closer than course mates as our days, nights, dining, entertainment, punishment, bathing, even dressing up is all done together in an almost collective manner. You may manage to hide your personality traits from course mates but it's impossible to do the same from your Squadron mates. It was a band of 25-odd brothers who underwent the training activities together for the next two and a half years

After passing out from NDA we landed in Singarh Company at IMA. The friendship bond established at Golf Squadron continued to be cemented further. We spent one year at IMA. On passing out from there both joined Infantry. Ajit was asked by our Company Commander then Maj DP Merchant to join his Battalion 18 MADRAS (Mysore). It showed that he identified the potential in him. Major DP Merchant went on to be promoted to a Major General before he retired. I joined 13 PUNJAB (Jind). Less than a year later we were together at Belgaum for our Young Officer's Course, most of the Golfies chose to be housed in one barrack. Ajit and I were roommates in a corner room meant to be shared by two officers where we underwent the four and half month's tough course

31

together. Ajit being a teetotaller used to have a sobering effect on me to ensure that I was well prepared for the next day's class.

Those were the days when internet and mobile etc were not present, so communication was mainly through letters or phone calls of Army lines; civil telephones were also not in vogue still we kept in touch. I got married in 1986 and was a father by 1988, and happened to go on Temporary Duty to Chennai. Ajit was posted as Instructor in OTA and was still a Bachelor met him there. I was glad that he had progressed to drinking beer and some hard drinks and was happy to have a bottle of chilled beer with him, that afternoon.

I was posted as DQ in a Brigade Headquarter at Dehradun in 1995. During my tenure there Ajit was posted to Army Headquarter in the MS Branch where he looked after the posting of officers up to the rank of Lt Cols. Ajit by now was married and father of two lovely kids Nirbhay and Akshay. He along with Shakunthala and kids visited us at Dehradun. It was he who gave me the good news that both of us had been approved to the rank of Lt Col, a week later he gave me another surprise that we had been nominated for the same Senior Command Course 50, from 8th December 97 to 7th March 98 which was to be held in Mhow. We both attended the course with our families and this time took our accommodation in the same block, my flat on the ground floor and his on the first floor. It is course which was quiet comfortable, with lot of time available to ourselves and we had a lot of family interactions also, it was fun to be enjoying the get together with limited resources and four brats in tow.

After the course, I got posted back to my own Battalion as the second in command whereas Ajit was posted to 25 RR with the same appointment. My Unit was in Kashmir Valley whereas his, was also in J&K in Doda area. Being in MS Branch he could have managed to get himself posted to the place of his choosing but he never tried that and banked fully on the fairness of the system. Life in J&K then and even now is quite challenging due to hostile neighbouring country's state sponsored terrorism. He was doing a magnificent job there, leading his troops from the front and in one encounter on 30 October 1999, sacrificing his life in the line of the duty to the motherland.

His untimely departure was a loss to his Battalion, The Madras Regiment and Indian Army as such because he had the potential to reach very senior ranks in the Indian Army. I have no hesitation to state what I did in the opening paragraph of this write up: I did not come across a more down to earth, fair, kind and sympathetic person than him in my entire career of 40 years in the Army. To us course mates, he was friendly, courteous, forgiving, helpful out of the way and fun to be with. His memory will remain forever etched in our minds till we meet him in Vallahala in our times to come.

Few words about Shakunthala, she's been an epitome of what we describe in the Army Parlance a 'Veer Nari', a brave woman who picked up the threads of her greatest personal tragedy in such an admirable manner that, not only she carved a promising career for herself but kept herself motivated enough to imbibe the qualities of her husband to their two adorable young sons. It's due to her mentoring that both Nirbhay and Akshay competed and qualified to join the Defence Forces, Nirbhay has joined his Dad's *paltan* 18 MADRAS (Mysore) and Akshay is a Naval Submariner now. This real-life story of the family is very difficult to imagine and is an example to the entire nation which can teach a lesson or two to everyone in motivation, sacrifice and patriotism. I wish them all the very best in their respective careers.

JAI HIND

Col Pradeep Katoch (Retd)

13 PUNJAB (JIND)

Play the Game

My school, Bangalore Military School, was earlier known as the King George School and had an English motto "Play the Game". The motto is from a poem by Sir Henry Newbolt, called "The Torch of Life". In the first verse, a cricket match is nearing its end, the players are exhausted, with just a few more runs to get, a win seems out of reach. The captain calls on his team to pull together, give a little bit more – "Play up! Play up! And Play the Game!". The backdrop of the second verse is the battlefield. The situation is dire, but as morale sags and men seem defeated, the cry goes up: Pull together! Once more! Play up and play the game!

The sentiments apply not only to the sports field and the battleground, they are also about the torch of life, the fire that keeps us going, the will that steers us through hard times, the strength that comes from working together.

Whilst these two quotes have helped me in my life on several occasions, the perfect embodiment and example of a person who followed the true deeper meaning of these quotes was my dearest friend Ajit Bhandarkar. Ajit and I were in different squadrons at NDA but at IMA we were in the same company Sinhgarh. In 1981, Sinhgarh Company was located near the stables in one corner of IMA and was famous for its languid and easy-going lifestyle. Needless to add that we came last in each and every event.

I distinctly remember one of the historic hockey matches between Sinhgarh Company and some marauding team from another company whose name I have forgotten. The match was part of the Inter Company Hockey Championship and Ajit Bhandarkar was playing centre half from our team. The cheering party of Sinhgarh Company was in full strength as the word had got around that the Company Commander Maj DP Merchant and all the platoon commanders would be present. The battalion commander was also expected.

The match started at 4 pm and within 15 minutes our company was down by a goal and each passing minute the chances of the score-board increasing exponentially was evident to all. By half time the score read 4 - 0. The performance of Sinhgarh Company was nothing much to write home about and all the players less one had given up the fight in the first 10 minutes itself. Whilst the rest of the Sinhgarh team was only going through the motions, Ajit Bhandarkar just did not give up and waged a lonely battle against a very superior team. He was simultaneously seen all over the field wherever the ball went, in the forward line at the 16 and 25-yard circles, defending in own half, ready for the penalty corners to help the goal keeper and virtually every where on the field. Whatever fight our team seemed to put up was solely the effort of Ajit. By half time it was clear to the winning team that they need to take on only Ajit and the rest would just cave in.

During the half time the entire hierarchy of Sinhgarh Company, including the Directing Staff and various appointments, did their best to infuse some josh and morale in the team but the sunken faces of the players and their curved backs gave some other indications. After lots of Glucon D and *Nimbu Paani* the players trooped in for the second half. But to the chagrin of the cheering party of Sinhgarh Company the story did not have a twist. The performance of the players now weary and tired after the first half chasing around was abysmal and they had lost all fight in them. However, our lone warrior Ajit had not given up and stood like a thorn in the winning teams efforts to increase the winning margin. Had Dhayan Chand seen him that day he would have been proud. I was amongst the lucky spectators to witness the single act of valour and courage put up by Ajit. He was a man possessed with nothing else but to play the game in the spirit it is to be played. He showed us that winning or losing is part of life but what matters ultimately is have you done your best, have you performed to the call of duty especially when the chips are down. That day he was the ultimate hero, the superman, someone whom everyone would want to emulate and ultimately a true officer gentleman in the true spirit of what IMA and Indian Army stood for.

The finale. At the end of the match as our team walked out with slouching shoulders and the familiar dejected look the Company Commander Maj DP Merchant called out only one player from the team - Ajit and addressed him in front of the entire gathering. Lauding his stellar efforts, he said his battalion 18 MADRAS (Mysore) would be proud and honoured if Ajit joined his *paltan*. Yes, that year-end Ajit got commissioned in 18 MADRAS (Mysore)!

As I look back at life and my career, I have finally understood that life is a game and we are mere players. But yes, the winner of this game is the one who overcomes all hurdles that come in his path and continues to play the game till the end like Ajit Bhandarkar.

Lt Gen Rajeev Sabherwal, PVSM, AVSM, VSM(Retd)

Corps of Signals

Chapter 5

Swadharme Nidhanam Shreyah -The Madras Regiment

श्रेयान्स्वधर्मो विगुणः परधर्मात्स्वनुष्ठितात् ।
स्वधर्मे निधनं श्रेयः परधर्मो भयावहः || 35||

śhreyān swa-dharmo viguṇaḥ para-dharmāt sv-anuṣhṭhitāt

swa-dharme nidhanaṁ śhreyaḥ para-dharmo bhayāvahaḥ

It is far better to perform one's natural prescribed duty, though tinged with faults, than to perform another's prescribed duty, though perfectly. In fact, it is preferable to die in the discharge of one's duty than to follow the path of another, which is fraught with danger.

The Madras Regiment, the oldest infantry regiment of Indian Army, is over 250 years old. The first twelve Battalions of the Regiment were raised between 1758 and 1766 to protect Coromandel and Malabar Sea coast. The troops of the Regiment have a rich tradition of martial valour and combat skills. They participated in the historic battle of Assaye and Bourbon (Mauritius). The Regiment has conducted itself with fortitude, bravery and elan in World War I and II. Post-Independence, the Regiment participated in 1947-48 J&K operation, in the Liberation of Goa, Indo-Pak conflicts of 1965 and 1971, operations in Sri Lanka and ongoing operations in Siachen Glacier. Also, the conduct and performances of its troops in Congo and Lebanon as a part of UN Peacekeeping missions has won them accolades.

On the merger of the states, the Mysore Infantry was integrated with the Madras Regimental Centre (MRC) of the Indian Army on 1st April 1951 and permanently merged with the Regiment on 16th April 1953. They were designated as the 18th Battalion, the Madras Regiment (Mysore) and call themselves the Mysore Marauders. The jawans of this Regiment are from Karnataka, especially from Coorg and Belgaum, and Kerala, Andhra and Telangana. This unit has accomplished with the Theatre Honour in the 1971 war, COAS unit citation for its achievement in OP Rhino in 1998, again COAS commendation for OP Parakram/ Rakshak in 2004 and the Force Commander Citation for the year 2010 and 2011, for its contribution to the United Nations Stabilisation Mission in Congo.

Ajit was successfully commissioned into the 18 MADRAS (Mysore) unit on 19th December 1981. The young officer undergoes an orientation training program at MRC and then joins the unit. In the 15 days of training, 2nd Lt Ajit Bhandarkar was introduced to the jawans' life, lived with them and also had a short crash course in Tamil, the regimental language. The Madras Regimental Centre is also a centre for training new recruits. The very well structured curriculum imparts all the facets of military training to a recruit thereby moulding him into a battle hardened jawan or a 'thambi' as they are colloquially called in the regiment. Subsequently on completion of training in the MRC, all the jawans are sent to the respective battalions they are assigned to.

The battalion was in Tamze, Sikkim, when the young 2nd Lt joined the 'Paltan' as is it fondly called. Having known the history of the Regiment, the War cry, "Veer Madrasi; Adi Kollu Adi Kollu" and the Motto "Nirbhaye Nirvikar", all members of the unit, start feeling the spirit of brotherhood and camaraderie.

As per protocol, all officers are dined in when they join the unit. A formal dinner is laid out with the Commanding Officer (CO), the Second in Command (2IC) and the other unit officers in the location. Col Chaman was the CO, Lt Col Ranga Raj was the 2IC, Adjutant was Maj OPS Pathania, and the other officers were Maj Sasi Kumar, Maj Sarkar, Maj Merchant and Capt Diljit Singh. Though Col Chaman was a teetotaller, the new officer, Second Lieutenant Ajit Vasudev

Bhandarkar was offered a drink. As the tradition dictates, a newly joined officer should drink at his dining in regardless of his likes or beliefs. Much to Ajit's dismay who happened to be a teetotaller then, just like his CO, had to uphold his battalion's tradition and started drinking until his neck started to drop and was slurring while he spoke.

By then his brother officer Col Diljit Singh (then Capt) very carefully lent his shoulders and both of them started walking towards their room. They had to pass a *nullah* for which they had to step on dry rocks to go to the other side. The minute he was told by Capt Diljit Singh, that the *Nullah* had to be crossed, he stood straight and upright. Ajit did not have issues with crossing the nala, which meant that Ajit to save himself from the senior officers had pretended to be drunk. I must admit, he had the presence of mind and displayed tact in a given situation.

PC: 18 MADRAS (Mysore) Records

Maj Gen DP Merchant AVSM (then Maj) to the left with Lt Ajit to the right.

Having served their tenure in Tamze, it was time for the unit to move. This time the unit was ordered to move to Secunderabad. Generally, the unit sends two to three officers and jawans as 'advance party' to the next station. The team makes all the arrangement for the arrival of the whole unit, which includes the offices renovation, accommodation for families, arranging the Mess and other formal initiation for the arrival, after which the rest of the *Paltan* moves.

So Ajit, with Maj Merchant, went to Secunderabad via Calcutta by train wherein only one bogie was given to the unit officers and jawans. They were to relieve 5 MADRAS. After reaching Secunderabad, Maj Merchant and Ajit shared the same room. Ajit was a little hesitant to share the room with his senior officer. However, Maj Merchant told him very affectionately that they were roommates and that he should make himself comfortable. Mrs Merchant also joined Maj Merchant in Secunderabad later and like always was the motherly figure; the lady's presence brought in a lot of cheer in the unit.

But there was something more serious: an order awaiting for 18 MM, which was a fresh order to go to Gandhinagar. So immediately the whole team had to shift and they all boarded a special train and went to Gandhinagar, Gujarat. Ajit was the officiating 2 IC, so he had to get the approval for the special train and many other permissions; which he carried out efficiently.

Finally, they reached Gandhi Nagar, with only two officers, i.e., Lt Ajit and Maj Merchant, a few JCOs, and about 30 ORs. At Gandhi Nagar, they had to take over a newly constructed accommodation. When the quarters were checked, there was a lot of plumbing and carpentry work which was either faulty or had to be repaired. So, all the initial teething problems had to be sorted by the escalating the matter to the Brigade commander and Station commander. Only after all the issues were sorted was the accommodation taken over. It was during this tenure when Ajit had to handle MES issues and Maj Merchant saw him at close quarters, his skills to get the work done from the babus and his relentless effort in maintaining high standards, may it be the food in the mess or the quarters handed over to the officers. All his seniors were in praise of him for his diligence and sincerity.

My First Meeting with My Senior Subaltern

I was undergoing training at Officers Training School, Madras (now OTA Chennai) in 1983. After rigorous training of five months, it was dream come true to proceed on the midterm break of 10 days. My parents were at Shillong, a beautiful hill station where I did my schooling and college prior to joining OTA. I boarded Coromandal Express from Chennai and reached Howrah Station the next morning. My connecting train Kamrup Express was in the evening, so had to spend a full day at Howrah before proceeding further. With the limited options that I had, I decided to go to the "*Fauji Aram Ghar*" and freshen up and then decide further course of action, to spend the rest of the day before boarding my next train.

I reached the *Aram Ghar* and was allotted a bed in the dormitory. Next to my bed was an Officer sleeping and seemed to be from an Infantry Battalion. A perfect *fauji* haircut, absolutely fit and sleeping like a typical *fauji* not bothered by the commotion around him created by commuters. I too decided to take a short nap before proceeding for lunch. When I woke up my neighbour was already awake. I quickly got up and we exchanged smiles. He then introduced himself as 2Lt Ajit Bhandarkar and I told him that I am a GC and going on a midterm break to Shillong, my hometown. He was proceeding on annual leave to Bangalore from Sikkim where the unit was. We got along very well and spent the rest of the day together. We had lunch together outside the Howrah Railway Station and came back to Aram Ghar as it was time for his train to Bangalore. Since my train was late in the evening, I accompanied him up to the train and we finally bid goodbye to each other with a hope to meet sometime during our Army life.

After completing my training at OTS Madras, I got commissioned into the Madras Regiment. By now I had forgotten meeting someone from Madras Regiment during my midterm break. Having done my attachment with the MRC, it was time to join my parent unit. The unit was still at Sikkim but the Advance Party had moved to Gandhinagar in Gujarat. I was told by Adjt MRC that I will have to join the Advance Party at Gandhinagar. Accordingly, the Advance Party was informed of my arrival date.

On reaching Ahmedabad, with great difficulty I was able to contact the advance party. The voice at the other end asked me to sit in the Waiting Room and that he will be coming within an hour to receive me. After about an hour there were two Thambis whom I could straightaway recognise because of the pom pom and one more young person who appeared to be the Officer. They were frantically looking for me. Since they didn't know me and I could identify them from their uniform, I walked up to them and told them "I am 2Lt Ishwinder Singh". The Officer looked very familiar but I could not talk much because it was my first day in the unit, so decided to remain quiet. I think the officer was also scratching his head to recollect if we had met earlier. It was only after about half an hour when I was little at ease we opened up and remembered our meeting at Aram Ghar, Howrah. Here was an officer who would remain my role model, my utmost regards for him, an ideal senior subaltern.

Col Ishwinder Singh Sehdeva (Retd)

18 MADRAS (Mysore)

CHAPTER 6

ENCOUNTER WITH THE VIP (BHARAT RATNA: SRI ATAL BIHARI VAJPAYEE) 18 MADRAS (MYSORE) IN GANDHINAGAR

All infantry units have a cycle of serving in "field stations" when they are deployed on LOC /LAC or are located at a border town for two years and after which they go to a "peace station", for the next three years. In peace stations, the families stay with the officers, and men also get an opportunity to get their families to the station. To maintain the same state of operational preparedness as in field stations, the unit conducts intensive training and participates in various military exercises even in peace stations thereby following the literal meaning of the adage - "the more you sweat in peace, the less you bleed in war". To keep the men motivated they have intra and inter battalion competitions like firing, drill, shooting and sports. So, to say, the army men are always ready and prepared to tackle any untoward situation.

Gandhinagar was supposed to be a peace station but then the unit was always involved in some operation or other. Either participating in military exercises or assisting local civil authorities when the internal security situation deteriorated. The CO and all the young officers stayed in rooms near the mess and dined in there. The unit had a busy schedule; apart from the usual military exercises conducted in the deserts, the period from 1983 to '86 was highly volatile.

Gandhinagar was in the thick of action. 1984 was an eventful year for India too with Operation Blue Star and the assassination of our then Prime Minister, Mrs Indira Gandhi. While the unit was all geared up for one operation after another, our young officers looked for relaxed moments whenever they would meet at the mess. The unit had talented singers and musical instrumentalists, but lacked a cohesive band with proper equipment. Some of the young soldiers had also got trained to play various instruments in Hyderabad and Secunderabad. All this pushed our young officers including Ajit to get a few instruments for the unit band, which otherwise used instruments made out of scrap metal and wood providing a cacophonous effect rather than a harmonious note to music. The young officers, after a formal sanction from the unit, procured the new set of equipment for the 18 MADRAS and named it 'Mysore Jazz Band' and thus the unit proudly had band performances for formal ceremonies and informal parties.

The Brigade Commander during one of the events saw the unit band and was so impressed that he ordered the band to perform for one of the events hosted in the deserts of Rajasthan. Now came the challenge and the tussle, the instruments were brand new and very delicate. Transporting and shifting them on road to the desert, exposure to the heat and dust of the desert would surely spoil the instruments. So, after the performance of the band for the Brig Commander's event, Capt Diljit and the then Lt Ajit were very much concerned for the Jazz band equipment and began cussing all the senior officers after a drink for using the unit's equipment to perform for an event in the desert. Now they were aware that the CO, Col Thanzawa would not support them on all these issues so both the young officers, late in the evening, walked a few miles, to meet the Brig, who lived in his caravan.

Since the band was very precious and valued by the young officers, both Capt Diljit and Lt Ajit, broke military protocol and mustered the courage to meet the Brigadier. They were very honest in their approach and admitted to their Brigade Commander that they had not discussed this concern with the CO and came straight to the Brigadier to voice their concern. Having very politely expressed their resentment in sharing and shifting their band equipment of and on, requested the Brigadier to keep their visit to his caravan confidential and not inform Col Thanzawa. Finally, they were successful in ensuring that the Jazz band equipment was safe and sound with them in the Unit location. Such incidents surely displayed the true regimental spirit of the officers and spoke volumes about their loyalty to the unit they served in.

The unit was ordered to move to Ahmedabad, for Internal Security (IS) duties. All companies of the unit were deployed in various posts at Ahmedabad, in Bapu Nagar, Walled City area and the Kanakria lake surroundings.

This was primarily because there was lot of protesting in and around Ahmedabad, to oppose the Reservation Bill. This commotion then turned into communal riots. As a young officer, Ajit was the Adjutant of the unit, and thereby, had to take a stock of the whole situation and give frequent updates to his CO and other seniors in the battalion. There was curfew imposed in Ahmedabad and movement of all vehicles restricted. An exempt pass was issued to all the vehicles and people who were allowed to move around during the curfew.

18 MADRAS (MYSORE MARAUDERS)

The Mysore Marauders, had been committed on Internal Security duties at Gujarat during the communal riots from March to July 86. The vigilance, alertness and zeal displayed by all ranks of 18 MADRAS (MYSORE) was appreciatd by both civil and army authorities. After such a trying, difficult but satisfying task, the battalion returned to the barracks,only to start a hectic training year.

We commenced our training cycle in time and were appreciated by the General Officer Commanding and the Commander for the excellent standards achieved. Our crowning event was the Inter Battalion Specialist platoon competition held in April 86. Out of the four competitions held in the Brigade, we stood first in Recoiless machine Gun and Rocket Launcher competitions.

Major BR Nayar qualified for the DSSC and is at present at our home grounds. Captain IS Sehdeva and Nb Sub Shamsudeen Kunju took part in a "Desert Safari", involving 350 km of enjoyable [?] camel ride.

Courtesy: The Black Pom Pom 1986 Edition, The Madras Regiment Magazine

The role of the Army Units was to check for these movements and ensure that the law and order is maintained. In the process, in one of the posts manned by 18 MADRAS (Mysore) unit, the jawans stopped a vehicle which had the vehicle pass, the driver too had his pass but the person sitting at the back did not have any pass. So immediately, they deboarded the person, who was a VIP and ordered him to stand with his hands up and face the wall. The soldiers were only following the orders given by their seniors. After sometime, the CO immediately reached the spot and recognised the VIP who was sweating profusely. When the CO enquired with his jawan, he promptly replied, *"Curfew pass nahi tha saab".* ("He did not have a curfew pass, Sir.") The jawans also failed to recognise the VIP, who later turned out to be the Prime Minister of India, Bharat Ratna: Sri Atal Bihari Vajpayee. Soon enough the Adjutant, Lt Ajit was called, who then very politely took the VIP with him and gave him a royal treatment in the officer's mess. Apologies were rendered to Sri Vajpayee and he was explained as to how an army unit functions. He was very graceful and brushed aside the apology for the misadventure.

After this tenure, Ajit had to attend the Battalion Support Weapon Course (O)-37 from February 87 to June 87. This is what Col Deepak Bhari had to narrate:

"YAARON KA YAAR"

It was pleasant weather in the month of February, in Mhow. We had come there to do our Battalions Support Weapons Course.

After finishing our RCL leg and the MMG leg, Guddu (Col Rajesh Tiwari), Ajit and I, had joined to do the Mortar leg which was longer than the other two.

We all had come well prepared to do well at this course and since we had spent some time in our respective units, we had a lot to talk about (crib). The best part was that the majority number of officers were from the 59 NDA Course (Infantry).

The routine during the Mortar leg was that during day time classes, we studied theory and evenings were spent cleaning the equipment in the *kote* (armoury) and getting familiar with it.

I remember an incident where we were told to practice hard for re-laying the Mortar to engage two targets simultaneously one after the other which required movement of the base plate to be measured by our hands (fingers spread) and levelling the bubble, to come into action immediately. We practised hard (the three musketeers), including the changing position amongst us.

On the day of competition, we all dressed in our smartest overall, presented ourselves for the test. First was Guddu to start followed by me and third being Ajit, in the sequence.

All went well till my turn, but when Ajit's turn came, I was to set the base plate and adjust the bubble to the middle in order to get the Mortar ready to fire for Ajit and Guddu was to align the site with the marker. Ajit was passing the grid references. All was set for the task. As soon as Ajit passed on the orders for shifting the base plate and adjust bubble to engage a new target, in exuberance, I forgot to shift the base plate by the required degree (using hand) and only just set the bubble in the middle. We all were so surprised that how come we did it so fast! Then we realised that I had bungled and had forgotten to move the base plate and we were still fixed on the first target. I was so ashamed of myself to have forgotten to shift the base plate as per the requirement and Ajit not able to score well on this test because of me. But Ajit, a big-hearted man, never ever mentioned this to anyone that because of me he did not do well in that test. He was such a magnanimous person.

Later, we laughed it out on my stupidity and had a chilled beer in the Officers Mess.

Ajit was always a man of friends - *"YAARON KA YAAR".*

Col Deepak Bhari (Retd)

Kumaon Regiment

Ajit as a kid standing extreme right, with his maternal grandfather

Ajit as an infant, lying down with his brothers Col Arun Bhandarkar (Retd) to the left and Mr Anil Bhandarkar to the right.

Ajit at NDA after a Camp, extreme left second row

Ajit during the Commando Course

'G' SQUADRON

59 G

SITTING (L TO R)   :   CQMS M KHARE, DCC P SAXENA, CSM PM CHERIAN, BCC V TIWATHIA,
                        3CC A KAPOOR, DCC M/S MANN, DCC 3 GOVINDRAJAN

STANDING (L TO R):   SGT KS AHLAWAT, SS JANEJA, 3 JAGGI, GK RAO, AS CHANDHOKE,
                        SGT PK KATOCH, AA KAMAL, CV RATNAKAR, V PRAKASH, RS GILL,
                        SGT A BHANDARKAR

STANDING (L TO R):   D BAHRI, SN ROY, R TIWARY, A MALHOTRA, VS BHARTI, VS RANDHAWA,
                        P. KAPOOR, P BISWAS, M KUMAR, DC PUJAY

Ajit at NDA - G Sqn Group Photograph

49

Ajit as a Young Officer

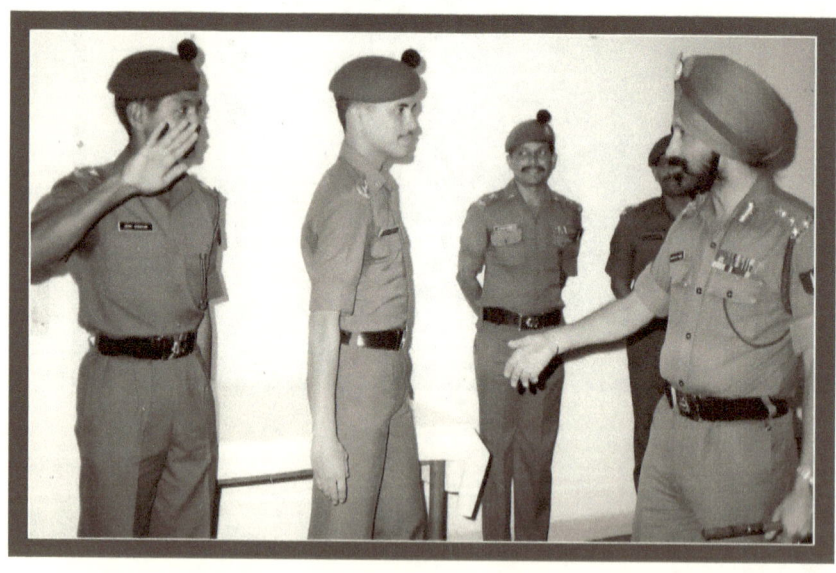

Ajit, second from left, as a newly commissioned Officer

Ajit (sitting extreme left) during Mountain Warfare course (42) in 1982

Ajit on his bike

OTA Trip to Ooty (Ajit standing second from left)

Ajit sitting behind the desk in uniform, with GCs of SS 48 course

Ajit as DS sitting in the centre amidst the GCs of SS 50 course

At Ferozepur, with CO Col Kuppa (Ajit sitting second from left)

Ajit as DS at OTA (standing second row, second from left)

Akshay's birthday celebration at New Delhi

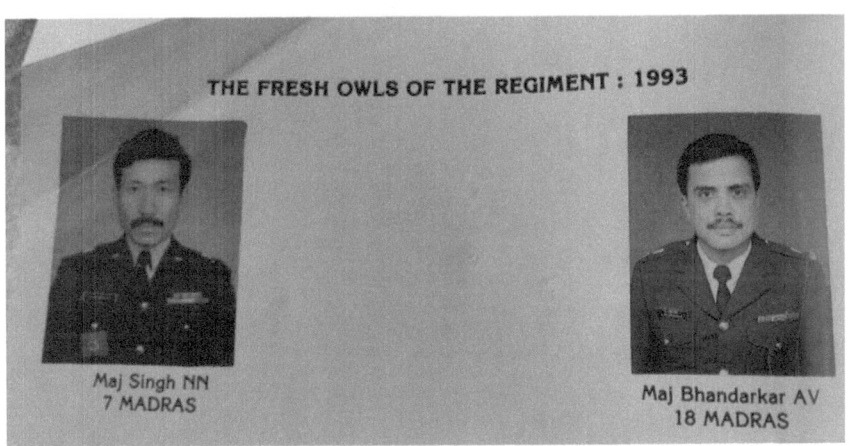

THE FRESH OWLS OF THE REGIMENT : 1993

Maj Singh NN
7 MADRAS

Maj Bhandarkar AV
18 MADRAS

Courtesy Black Pom Pom - 1993, Biennal Number

With Unit Officers at Jamnagar (second from left)

Ajit's favourite way of carrying his boys

Junior Command Course Ser 53 – Aug 1988 (Standing second row, third from right)

57

Senior Command Course 80 – Mar 1998 (Standing last row, sixth from left)

Ajit wearing a beanie cap, standing at the back, to the right of his CO, Col Ramachandran

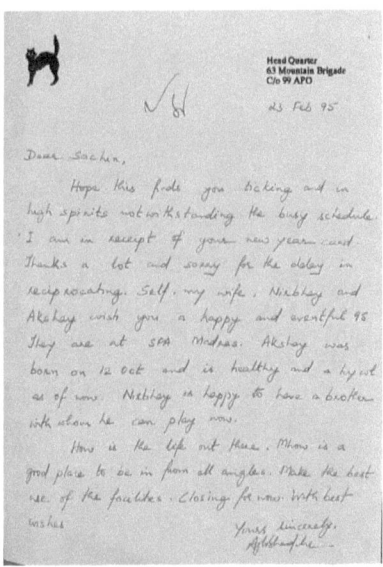

Ajit's Letter to Col Sachin, his student in OTA

Ajit with his hands at the back, briefing his team

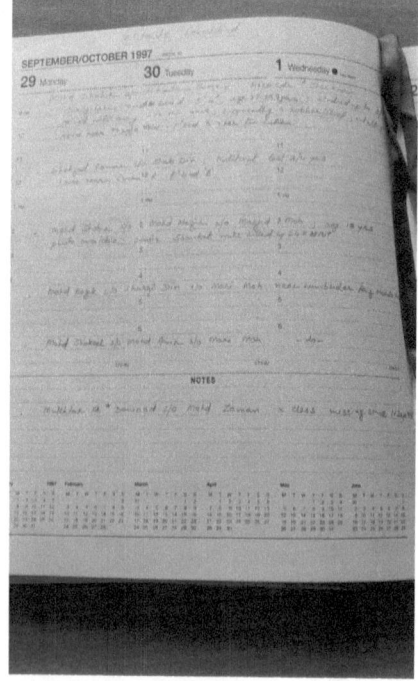

Ajit's personal diary containing details of militants

Bhandarkars family portrait

Ajit Dwar at Sainik School Bijapur

Mrs Shakunthala receiving the Saurya Chakra Award from the President

CHAPTER 7

SERVE WITH HONOUR – OFFICERS TRAINING ACADEMY: AN INSTRUCTOR WHO TOUCHED THE HEARTS OF HIS GC

The history of Officer's Training Academy transpires like this:- Following the Indo-Sino war in 1962, there was a need to train more officers into the Defence Services. So, the Officers' Training School, as it was previously called, was formed. The process to begin this school started in September 1962 and was inaugurated on 15 January 1963, with Brigadier Ram Singh as its first Commandant. The OTS, which later got its name Officer's Training Academy, continued to operate and on 2 February 1965, obtained the sanction to shift focus to train officers for the Short Service Regular Commission.

True to the motto of this great institution, 'Serve with Honour' this place has produced many war heroes and distinguished officers, who after serving the Armed Forces, have joined other organisations and have made a name for themselves in several fields.

For the entry into Officer's Training Academy, the candidates should have completed their graduation and should clear the Combined Defence Service (CDS) Exam which is conducted twice a year. Only unmarried graduates are eligible to appear for the exam. Successful candidates are admitted into the respective Academies after an interview conducted by the Services Selection Board (SSB).

In the Academy, all trainees are called Gentlemen Cadets (GCs). They are organised into six companies: Meiktila, Naushera, Kohima, Jessami, Zojila and Phillora. The training here is both physically and mentally excruciating. The training in the academy truly redefines the quality of the GCs. His body, mind and spirit are toughened so that they are capable of facing any danger or fear with confidence and courage. The training standards are time tested and in sync with the military traditions and groom all the gentlemen cadets and lady cadets to become great leaders of tomorrow.

Ajit was posted to OTA as an Instructor on 26 November 1987 and was in charge of the Kohima company. As an instructor, he was a role model to many of his GCs. He would frequently have an informal chat with them and get to understand them. There were a few cadets who had come from a civilian background and had no knowledge about army life and its tough training norms. They did feel intimidated by the strict regime and the exacting routine. He would, like a good friend, explain to them the purpose behind the rigorous training schedule.

Capt Ajit V Bhandarkar taking a class

The young cadets were fresh from college and away from their homes. The arduous training and jam-packed schedule of outdoor camps, inter-company competition and the pressure to perform their best would sometimes break even the toughest person. As a Directing Staff (DS), Ajit closely monitored and mentored each and every one like a father figure. He would closely keep an eye on the GCs during the practice sessions, accompany them on their outdoor training and camps, and motivate each and everyone in the Kohima company to do their best. His company was adjudged the best passing out company of 48th SSC course.

It was during Ajit's tenure in OTA, when his parents had come to visit him in Chennai, that they insisted he get married and start his family life. He had enough fun as a bachelor and it was time for him to be serious with life and take the plunge of getting married.

Capt Ajit Bhandarkar with his GCs

My father was working in Chennai and I was in Chennai too, working as a school teacher in a nearby school. It was through one of our family friends that we were introduced to Ajit and his family. My dad's cousins and friends questioned him; why did he want to get his only daughter married to an Army officer; he very proudly said that it would be his honour to have a son-in-law serving in the Indian Army. I, for one, had a vague idea of Army life after watching the serial *Fauji*, on Doordarshan enacted by Shahrukh Khan and remember the crush my college friends had on Army officers, while I was studying in Ajmer.

After having a personal interaction with Ajit for an hour, at his cousin's place, I was completed floored by Ajit's sincerity, in explaining to me, without mincing any words, about the hardships and challenges I would face if I married him. He proved himself to be a thorough gentleman, an honest guy, worthy of being my husband. So, after Ajit's acceptance, our courtship started.

One day while I was busy taking my class in school, the admin clerk came running to me, saying that there was a call for me. In those days, it was landline connection. I had to go to the office and take the call. Ajit called me to ask if I could join him for lunch. I had to say no to him saying that I was at work and couldn't leave my class abruptly. After which we fixed an evening date.

Ajit came home and took me on his blue Kawasaki bike to the Academy. By the way, blue was Ajit's favourite colour. All his T-shirts were either blue or brown in colour till the time we got married. As we entered the campus, the sentries gave Ajit a crisp salute and I was instantly impressed with the aura surrounding the Academy. I was entering an Army cantonment for the first time and that too a prestigious institution like the OTA where thousands of officers have been groomed and passed out through the hallowed portals successfully. I met his colleague, course mate and squadron mate and mischievious as always Capt (now Col Retd.) Vikram Tiwatia. After spending some time with Ajit and Capt Vikram, I was dropped back home.

Now my colleagues in school, where I was working, came to know about Ajit as my fiancé. They would start enquiring about our meetings and conversations with him. One friend of mine started pulling my leg by asking me personal questions, like "did he kiss you or did he hug you?" When I told them that it was only meeting and chatting, they would not believe me! I truly like the respect and the space he gave me to understand him during the few months of courtship before the marriage on 11 October 1990.

REMINISCING THE DS, CAPT AJIT V BHANDARKAR

I qualified for SS-49 Course and joined Officers' Training Academy (OTA) at Chennai on 5 May 1989. The routine at OTA was strict and we hardly got time for ourselves. We used to have lots of fall ins where we were toughened, that is a decent way of addressing ragra at OTA. If we were spared by the seniors the Directing Staff used to take us on. There was not a single day when we did not have ragra. We often used to discuss our respective DS with cadets of other companies. It appeared that Lt Col Ajit Bhandarkar (Capt then) of Kohima Company was a thorough gentleman, who was adored and idolised by his cadets. His smile was sweet yet curt. We all cadets wished that we were in his Company. Slowly days passed by and we were due for our midterm outing around December 89.

Bangalore, Hyderabad and Ooty were selected for outing and the cadets were asked to give their choice. I had volunteered for Ooty and to my joy, the overall responsibility for conducting the trip to Ooty was given to Capt Ajit Bhandarkar as he was from Madras Regiment. All the cadets travelled by train from Chennai (Madras then) to Mettupalayam for our onward journey to Wellington and Ooty. On reaching Wellington we were made to stay in the barracks of Madras Regiment Centre.

The next day we visited the Centre and the Officers' Mess and did some local site seeing. On the third day, we were taken to Ooty and Conoor and Capt Ajit Bhandarkar accompanied us to both the places. We had a wonderful time here. Not even for a moment, we got the feeling that our DS was travelling with us or he was different from us. His demeanour was more like us. After seven years of commissioning, I was posted to HQ, NSG and came to know from a common friend that Lt Col Ajit Bhandarkar was posted at Army HQ. I immediately picked up the phone and spoke to him at length. To my surprise, he remembered me and my brother Brigadier Pankaj Dimri and also our visit to Ooty, even after so many years.

After a few months, Lt Col Ajit Bhandarkar got posted to Rashtriya Rifles Battalion in the valley and then one day we learnt about his martyrdom in the valley. It all came as a rude shock to me because only a few months back I had such a hearty conversation with him.

With Lt Col Ajit Bhandarkar attaining martyrdom, Indian Army not only lost a gallant officer but also a very upright gentleman.

Col Vikas Dimri

21 Jat Regiment

49 SSB

Kohima 16 DS: SS 48

I am from the 48th SS course and Capt Ajit Bhandarkar was our DS. I was the Senior Under Officer, (SUO) in the second term and would report to Ajit sir every day. We were blessed to have Ajit sir as our mentor. He was liked by one and all because there was something special in him. He was able to connect to youngsters and bring all of us together, in the Military academy, where discipline is driven into everyone by force.

It goes without saying that the professional competence of Capt Ajit Bhandarkar was outstanding, and his posting as an instructor in an Academy which trains gentlemen to become officers and lead men was a testament to this. He was the best of the best!

We, as a platoon, were slowly formed in the absence of our platoon commander. Ajit sir joined us a bit late and our seniors would always warn us and state when Capt Ajit Bhandarkar comes you will have tough times. Soon the command was taken over by the Tiger (Ajit Sir). We had a perception of him being strict by all standards, but his compassionate nature and his approach as a friend, philosopher and guide was slowly overpowering all of us. He cared for each and every cadet under his command and at times went beyond the call of duty. These qualities of his were slowly sinking in all of us.

So, our average platoon was slowly and steadily being trained in a manner which was not only as per the military norms but was also being coached in a motivating manner. His humane values overpowered us to the extent that Kohima company of SS-48 course was adjudged the best company under the leadership of Capt Ajit V Bhandarkar.

The company commander Maj Sampat has moved out and now the additional responsibility was taken over by our platoon commander. So, it was not only our platoon but two others which were being looked after by Capt Ajit Bhandarkar.

It is not this banner which was awarded to the best company but the manner in which he trained us in OTA that we owe so much to him. As the saying goes "100 sheep under one tiger will behave like a tiger and 100 tigers under one sheep will behave like sheep".

Yes! He was a tiger and he made many more tigers not only in the platoon but way beyond: a true brother, father and a friend. He made officers out of the gentlemen cadets in a way that whole platoon of Kohima-16 (SS-48) owes a lot to him.

Col P K Daka

Director Works

Ajit , My School Mate

Ajit was three years junior to me. I knew him more because of my cousin, Late Ravi Chengappa, Roll No. 547, Chalukya House.

My recollection of Ajit from school was that he was a quiet, confident boy. Good in studies. Later we stayed with him for a week or so in April 1989 when I had appeared in the Staff College Entrance Exam in OTA. My wife, who was expecting, had joined me. We stayed in the room next to Ajit and spent some time with him. Ajit lent us his motorbike for our jaunts around Chennai. Ajit then came across as a large-hearted and jovial friend. He was focused in the execution of his responsibilities as an instructor in OTA.

Major General Arjun Muthanna, (Retd)

Roll no. 220, Sainik School Bijapur

CHAPTER 8

SULAGNE SAVADHANA – MARRIAGE IN CHENNAI

My first meeting with Ajit happened at Chennai, in January 1990. By March '90 Ajit was posted back to the unit which was located in Ferozepur, Punjab. Though he was supposed to come to Chennai in May for the engagement ceremony, because of various commitments and also due to some operations happening in the Brigade, across the border, he could not come down south. After confirmation from Ajit's parents, the wedding date was fixed on 11 October 1990 and the formal engagement happened on 10th October when we exchanged rings, which was worn by Ajit for only a week after our marriage. He felt all these accessories like a wedding ring, or a chain would be a deterrent while executing outdoor tasks like firing, cross-country runs, or even for that matter while playing football or hockey.

Chennai in the month of October is pleasant and an ideal time to flaunt the Kancheevaram and Banaras sarees. In early September, my parents and I drove down to Kanchipuram not only to buy my wedding sarees but also to buy wedding gift sarees for Ajit's mother, my sister-in-laws and other aunts. My wedding sarees were specially designed by me which the weaver took a fortnight to complete and deliver it to us. In our custom, the bride, on the day of the marriage changes her saree a minimum of three times. For all important ceremonies during the marriage, the bride changes not only her sarees but jewellery too, which were all specially handpicked and made. In fact, my dad got the jeweller to come home and he embedded the diamonds in the gold earrings, in the conventional pattern. This was to ensure the quality of the setting and also to ensure the diamond

70

was the original one which we had given. So to say, a lot of effort and money was spent on sarees and jewellery.

Even when it came to the food served during the wedding, the raw materials for all the various meals and snacks were purchased by the host and given to the head cook. The head cook would then prepare the dishes according to the menu and the number of guests. It was unlike the present times which is less complicated, where we either have an event manager taking care of all the arrangements or a contract is given to a catering agency for all the meals for the wedding.

Keeping in mind Ajit's feasibility and his opinion, everything right from the invitation card to arrangements in the marriage hall was made. In the invite, Ajit did not want his rank to be displayed and also for his wedding dress he did not want any flashy silky wedding sherwani and the fancy turban or the "*Mysore Peta*". He said he wanted to wear "*Gandhi Topi*" and a simple cotton *sherwani*. Basically, he wanted a simple wedding and in fact, when my father wanted to discuss all the arrangements with him, he insisted on having an undemanding wedding. I was slowly getting to understand his taste and attitude which reflected his personality of being humble and unassuming. But for my dad, because it was the first marriage in the family and I being his only daughter, he wanted to have an extravagant wedding. So, it was a two-day wedding affair, on the first day the welcoming of the groom and the "*Var Pooja*"[1] (felicitating the groom) followed by the "*Phool Mudi*"[2] and the main wedding ceremony on the second day.

On 10th October, Ajit, along with his parents and all his family members arrived in Chennai by train while a few amongst them drove down. Accordingly, my father had made all the necessary arrangements to ensure a comfortable stay for all the guests attending our wedding. Many of our relatives,who had come from overseas too were gracious enough to be present and shower us with their blessings. On this day, we had the ring ceremony wherein we exchanged rings after which we interacted and were introduced to the guests.

---

1   Var pooja: felicitating the groom

2   Phool Mudi: phool means flowers which will be adorned by the bride. Mudi means ring, which will be worn by the bridegroom

On Thursday, 11th October 1990, our wedding day, we had many ceremonies according to our Gowda Saraswath Brahmin's custom. As a bride, I had to perform a line of rituals, starting from offerings to the various elements on Earth, the priest was eliciting the significance and its relevance to us. After the *Mantap pooja*[3], *Kasi Yatra* and other ceremonies we took the sacred vow of marriage, walked around the holy fire to religiously complete the wedding proceedings.

Our wedding ceremony

After which the mother-in-law bestows a new name to her daughter-in-law. In my case, since both my mother-in-law and I share the same name, I was named as Namratha. We also played a couple of traditional games after the marriage, like finding the ring from a vessel filled with water, rose petals, coins and a ring. The belief is if the groom gets the ring, then he would dominate over his wife and if the wife finds the ring then she would rule her husband. Between the two of us, Ajit found the ring. Many of Ajit's friends had come from OTA and also, we had one serving unit officer, Capt Hundal, who represented the 18MM at the wedding.

Ajit's parents, having come from Bangalore, immediately returned. Ajit and I stayed in Chennai to complete a couple of formalities as a newly married couple which included visiting the

---

3 Mantap: the holy place under which we get married

temple and also my first visit to my parent's house as a newly married woman.

After the wedding, feeding each other

The same weekend, i.e., on Sunday the 14th of October 1990 we had a *Satyanarayana Pooja* combined with a reception in Bangalore, where I met all of Ajit's friends and relatives. After which it was a chain of a continuous visit to all his cousin's homes one after the other. While Ajit had planned our honeymoon and we were all set to go to the Andaman Islands, I got the shocking news that my dad had met with an accident while he was going for his morning walk.

So, the rest of the holiday was spent with my parents in Chennai, because my dad was seriously injured and had to get his left leg amputated because of the wound. Though my dad was still recovering in the hospital, my brothers insisted that I accompany Ajit to Ferozepur, assuring me that they would take care of him. So Ajit and I together reached Ferozepur in November 1990.

Ajit Bhavaji[4]: My Dear Sister's Husband

Ajit *Bhavaji,* was a man of few words. His actions, instead, spoke louder than words. Unfortunately, I had just a few opportunities to spend time with him due to our work and location constraints. Yet,

---

4 Bhavaji: Konkani term for brother-in-law

those few instances gave me an inspiring glimpse into the amazing man that he was.

I clearly remember a few incidents that showed how down-to-earth and kind a human being he was. In late 1990, just a few days after Bhavaji married my sister Shaku, my father met with an unfortunate road accident and had to be hospitalized. I recall Bhavaji cancelled the honeymoon plans and stayed back to help us deal with that unexpected and stressful situation. On another occasion, during one of his stays at our Chennai home, he quietly cleaned up one of the bathrooms without any fuss.

He was an avid reader and every time he would visit us, he would read *The Hindu* newspaper from cover to cover. He wrote well and I remember his beautiful cursive handwriting in the letters he would write to Shaku and us. His handwriting, like a perfectly aligned row of soldiers marching in line, was neat, upright and graceful. He was a music-lover with a well-curated music cassette collection that was neatly labelled by hand also reflecting his simple, humble and straightforward nature.

He was also a well-dressed man. Simple and elegant. Nothing showy. I fondly remember his first gift to me was this lovely set of a branded grey shirt and pants. They fit me so well that they remained my favorite for a long time.

Bhavaji attended my wedding in 1996 when I saw him last. But the last time I spent quality time with him was in Wellington before I left for my US job in 1993. I remember chatting with *Bhavaji* in the evenings while enjoying a glass of rum with him. How I wish I could do that today.

Mr Narendra Kamath

IT Professional based in Canada

CHAPTER 9

UNIT LIFE IN FEROZEPUR

My first visit to Punjab was also the first time I was crossing the northernmost part beyond Delhi. The warmth and affection shown by both Ajit's brother officers and ladies were truly overwhelming. We were received by Col Benny Sebastian (then Capt Benny Sebastian), with flowers at the station and also a hot cup of coffee and some biscuits following which we proceeded towards the Unit mess, where we stayed till, we got our accommodation.

We had travelled by train from Bangalore to Punjab. Throughout the journey, Ajit was reading some book, listening to music on his Walkman and also talking to me, whenever we sat to have a meal/snack. He was counselling me about my roles and responsibilities as an officer's wife – the family welfare meets, ladies club, the mess etiquettes and the protocol I had to follow with the senior officers. He was very clear from day one that along with him, I need to accept the organisation too, which he was very proud of.

Outdoor campsite visits: Ajit to the extreme left.

During the initial days, we went around calling on all the officers, the experience was very new to me. He meticulously scheduled all the visits to unit officers' houses. He was observing my behaviour too with other ladies. On one such visit, I refused to have any drink, soft or hard. Said no to hot beverages too in the month of December, since I was not used to drinking tea or coffee down south. I was one of those goodie-goodie types who had only milk or tea only once in the evening. The next day, while we were having our breakfast, Ajit told me that it was very rude to say "no" to any of the drinks offered by the host. So, from then on, I began to drink tea/coffee during any meet or visit.

Ajit, sitting to the left with his brother officers of 18 MADRAS (Mysore)

Gradually, I was getting accustomed to Army life in Ferozepur. The officers would go on exercises and we ladies would spend time together. I would wholeheartedly prepare an assortment of snacks like *chakli*, diamond chips, etc. and send it to Ajit, in the one-ton vehicle which delivered ration and material to the officers and jawans deployed at different posts in operational area.

One fine day Ajit came home and he was accompanied by junior officers. I was very happy to see him back, while I was still a blushing bride, shy to look into Ajit's eyes in front of his colleagues. It was then that one of the officers told me, "Ma'am look at sir!" I then lifted my head and looked at Ajit and smiled. Once again, the officer said, "Ma'am look at sir properly…". I didn't get the reason why the officer was repeatedly asking me to look at Ajit. It was only after he told me that Ajit was promoted to the rank of "Major" that I saw the four lions in *ashok stambh* insignia on his shoulders. How can one forget these precious moments!

A group pic of 18 MADRAS (Mysore) unit officers: Ajit, standing third from left

While I got the hang of the tradition and culture in the Unit, Ajit was supposed to join the three-month Intelligence Course in Pune. So, there we went packing our bags and baggage and landed at the Military Intelligence School, Pune in April 1991. It was here I saw the way families lived when the officer was doing the course. As young course participants, we were allotted a temporary accommodation for three months. We set up our household and were ready to roll out meals not just for the family but sometimes for a dozen bachelor officers, who were doing the course.

Ajit, being a serious soldier, was very focused on his work and course while I would be busy cooking, taking care of the laundry and other household chores. Occasionally we did go driving to the suburbs of Pune like Alibag, a short sight-seeing trip to Sinhgad fort and a tour of the National Defence Academy.

Ajit, standing extreme left with some of the foreign officers

Military Intelligence School - Course Ser (56) Report

After completion of this course, we went back to the Unit. Ajit, true to his personality, was very focused on his career and his duty. He started talking about the profession, as an army officer, how they had to continuously upgrade themselves, stay fit and continuously keep reading and clearing various exams. This is when Ajit started preparing for the Defence Service Staff Course (DSSC). In 1991, we went on leave for only two weeks to our parents' place in the south and were back to Ferozepur after the leave. Ajit utilized his annual leave to prepare for the DSSC exam. DSSC exam is a competitive exam for all the three services, Army, Navy and Air Force for which thousands of officers appear and only around 200 get selected.

So Ajit and I stayed in Ferozepur during his annual leave, while he was busy and seriously preparing for the exam. He had a meticulous routine, wherein he spent almost 10-12 hours a day studying, making notes and reading the newspaper.

At this stage, during our casual conversations, he would always mention that he was very fortunate to get his annual leave for preparing for his exam and it was very thoughtful of the CO, then Col G. Atmanathan, (now Brig Atmanathan, Retd) to approve his furlough (leave without pay). He was very committed towards his goal of clearing the DSSC Examination.

In the meantime, he would also talk to me, about the level of toughness in these competitive tests and how clearing it was difficult and only the topmost scorer gets selected. At every stage, he would prepare me for failure. This is how, slowly and gradually he was mentoring me to tackle a situation thrown at me.

One day, during the summer of 1992 we got to know that Ajit had not only cleared the competitive exam, but was also going to become a father. This good news was shared with all our near and dear ones. And when I was in the 3rd trimester, I moved to my parents' house for prenatal and post-natal care.

CHAPTER 10

YUDHAM PRAGYA- DEFENCE SERVICE STAFF COLLEGE, WELLINGTON

Ajit continued to serve in Ferozepur, when I was down in Chennai with my parents, where my mom was feeding me with all the traditional dishes an expectant mother should have, to my heart's content! Generally, it is custom in the south that the daughters go to their parents' house for the delivery of the baby. We had a short and sweet baby shower at home with our family and friends, including my parents-in-law and all the elders at home.

Ajit was granted leave for a few days before the due date of delivery, so he was with me on 5th November 1992 when Nirbhay was born and was present during the naming ceremony too which was held on the 11th day after his birth. While I continued to stay with my parents, Ajit went back to the Unit in Ferozepur. I felt grateful that God had been kind and blessed me with the most loving husband, who not only cared for me but also made me the strong person that I am today.

It was during the summer of 1993 that Ajit packed all the household stuff from Ferozepur and sent it via road to Wellington. He came down to Chennai to pick me and our 6-month-old baby, and we headed to Wellington. I must admit the way the household items were packed were impeccable, all the trunks were duly numbered and were marked as "essentials", "linens", "drawing room items" and their list stuck inside the trunks for reference. Therefore, it was very easy for me to unpack and start my kitchen at Wellington. I am sure all my army friends will agree that this is how service personnel tackle the frequent transfers.

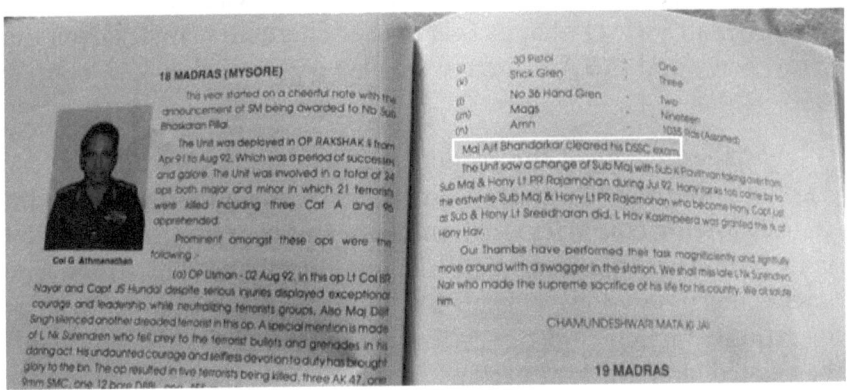

The 18 MM update in the Black Pom Pom 1992 edition informing Ajit clearing the DSSC exam.

The Defence Service Staff College, Wellington, Tamil Nadu, is a joint services institution, where officers from Army, Navy and Air Force and those from friendly foreign countries come for training. The aim of the Staff Course is to train selected officers of the three services in command and staff functions in peace and war in their own service, inter-service and joint-service environment, and provide related general education to enable them to perform effectively in command and staff appointments.

Another important establishment, in Wellington, is the Madras Regimental Centre. This centre trains young lads from Tamil Nadu, Kerala, Karnataka, Andhra Pradesh, and Telangana to join the units of the regiment. They are very young and enthusiastic youths from the rural region and suburbs who are passionate about joining the army. The centre trains the other ranks and prepares them for the tough life in the army that is in store for them. The training includes firing, drill, weaponry, fieldcraft, and other skills required in their service career. Above all, they are taught to operate as a team and traits such as camamaraderie and courage are engrained into them which are paramount while operating in war-like situations.

With our little son, Nirbhay, we were eagerly looking forward to the tenure at Wellington, for many reasons. First of all, it was one of the most prestigious courses for any young army officer to attend. Secondly, it happened to be very close to our home station i.e., Bangalore where Ajit's parents were and Chennai where my parents were. Thirdly, the Madras Regimental Centre was also in Wellington. Fourthly, Ajit was very happy that many of his course mates from the Army, Navy and the Air Force were also a part of this 49th course, 1993-94.

For the benefit of the readers, this course is conducted every year with almost 480 officers from all services, and 20 officers from friendly foreign countries also joining in. This course is for a duration of 10 months with a few breaks during Diwali and Christmas. The officers discuss various subjects like military history, the conduct of operations, tactics, logistics, science and military technology and the related staff work at various headquarters. They also go on an educational tour to various institutions to understand the workings of various defence establishments.

Ajit remained busy with his course. He had already warned me that he would not have any time for household chores and family. Ajit had borrowed his father's Fiat and we had driven to Wellington. So once in ten days we would go to the grocery shop at the Madras Regimental Centre (MRC).

While Ajit was busy during the course and preparing for his presentations, I was engaged with Nirbhay. I had started him on solid food, so I would cook and mash his food for lunch; give him a good massage before his bath and then take him out for a walk to bask in the mid-morning glow of the sun, which was a warm treat in the not so sunny Wellington. Apart from this, since it was very close to our home town, we had a stream of relatives visiting us every now and then. As Ooty is a favourite tourist spot in Tamil Nadu, many of our cousins made it a point to visit us.

It was during this tenure, that I realised, how I had to learn to manage all by myself, in spite of the fact that I was with Ajit. It was somewhere in November, when it was very cold and nippy at Wellington, that my dear son, Nirbhay, fell ill and almost didn't wake up from his afternoon nap. Since he was an active kid, he would not sleep in the afternoon for more than two hours. He would wake up promptly to have his food and milk. That day even after three hours,

Nirbhay, who had just completed a year, did not wake up. I went to see him in his bedroom and when I touched him, I realised that he had high fever. Ajit was busy with his course; there was some important session so he could not come home immediately. So, I had to take Nirbhay to the Military Hospital where he was admitted for observation and medication. Ajit came to the hospital at 8 pm after his meeting, when everything was under control. The paediatrician had given him a dose of paracetamol and after examining told me that he had heavy chest congestion. It was the first time that I was dealing with an emergency situation all by myself and was in tears. So from that day onwards, I understood, that for Ajit, his duty towards the organisation would come first before family.

Ajit sitting down, with all his school mates and their families.

During the course, student officers go on a couple of educational tours to visit premier defence establishments of all the services and get first-hand information about inter-services synergy and functioning. Finally, after submitting his dissertation, "Nuclear Environment in Southern Asia, Nuclear Non Proliferation Treaty and India's Nuclear Option", Ajit was conferred with an M.Sc in Defence studies.

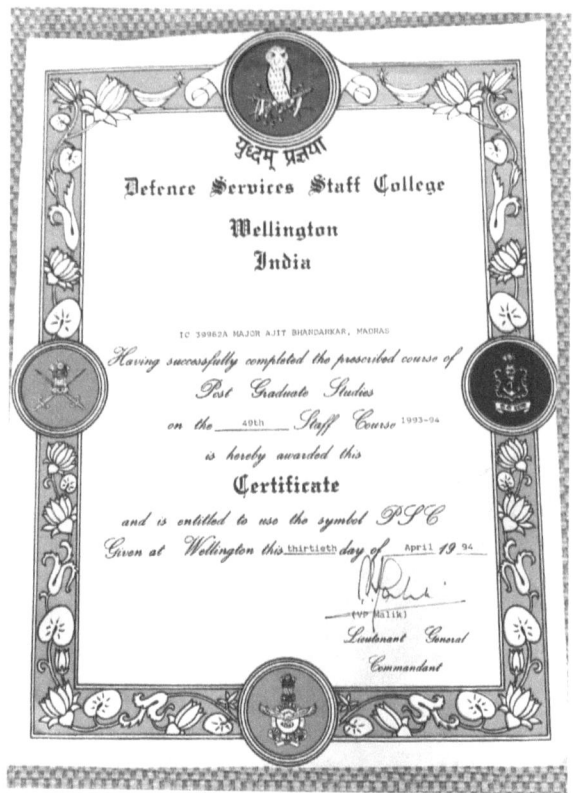

Certificate from DSSC

Apart from the regular AWWA sessions, we had for ladies at DSSC, as a Madras regiment officer's wife I had to contribute to welfare of the families in MRC and attend to meets organised by the ladies' club. The regular visits to the Gymkhana Club for a get together or a Sainik School meet were thoroughly enjoyed. The scenic beauty and the weather in Wellington were salubrious and time flew fast enough to say goodbye to Wellington.

After completing this course, Ajit was posted as a Brigade Major of a Mountain Brigade which was deployed on the historic silk route: the Nathu La Pass on the India China Border.

AN ANGEL IN THE TWILIGHT

It was during the monsoons of 1993. Ajit and I were attending the Staff College at Wellington. The salubrious climate of the Nilgiris beckoned the students to don their Sunday best, re-charge their wicker baskets with goodies, scoop their cackling toddlers and head out to Ooty; to unwind from the drudgery of voluminous and repetitive military appreciations and mundane wargames.

It was during one of these sojourns that my wife, Renu, six-month-old daughter and I, during the return trip to Wellington, were stranded about 10 kms from home. The rain God was unrelenting. The Maruti 800 decided to take in copious amount of water under the engine hood and went into slumber. Standing under a tin shed on the side of the highway, trying to shield the baby and waving our arms like windmills, at every passing vehicle, I am sanguine we were a poor sight to any onlooker.

A good old Fiat swooshed past us and turned the hair-pin bend towards Wellington. We could barely make out the occupants in the twilight. As our despair was turning into a mini panic, I saw the car coming towards us. It was Ajit who had barely recognized us in the fading light and had returned to confirm and rescue us from our plight. Shakunthala and young Nirbhay made space for us in their vehicle and forever in our heart. Truly, Ajit was an angel in the twilight.

Brig Vikas Puri (Retd)
Army Aviation

LIE DOGGO

It was the second week of November '93. Wargame and Divisional Sand Model Discussions (SMD) had taken its toll. Diwali and long weekend were around the corner. Ajit and I were discussing our plans (no, not the SMD but weekend ones). Not wanting to be Guderian or Patton in the SMD, both Ajit and I decided to "lie doggo" for the four-hour torture that was inevitable. We were oblivious of the fact that we were in the cross-hairs of our instructor, Col (later

Maj Gen) RS Mehta. During the tea-break, Col Mehta stretched his 6-foot-plus frame and looked down at Ajit and I and declared that if we did not open our mouth and take "active" part in the SMD, he would be constrained to send us to MH to check if we suffered from Lock-Jaw. Needless to add that Col Mehta was not smiling when he said that.

Post the tea break D Division was surprised at the sudden "josh" and the knowledge eruption that was forthcoming from the two of us. We were as if men possessed! From the viewer's gallery Col Mehta nodded his appreciation. I am sure many in the D Division were bewildered by the sudden and continuous firepower from the two of us.

As part of the Editorial Team of DSSC publication "OWL", I was reminded of this incident and decided to include this in the Course-end pen picture of Ajit.

"He was like a cork, surfaced every now and then - to say his piece and lie doggo once again."

Brig Vikas Puri (Retd)
Army Aviation

CHAPTER 11

BRIGADE MAJOR AT NATHU LA

Nathu means "listening ears" and La means "pass" in Tibetan. This Himalayan mountain pass in East Sikkim district is 14,000 feet above sea level and is 54 km from Gangtok. Nathu La is one of the three open trading border posts between China and India, the other two being Lipulekh, Uttarakhand and Shipkila, Himachal Pradesh.

Sealed by India after the 1962 Sino-Indian War, Nathu La was re-opened in 2006 following numerous bilateral trade agreements. The opening of the pass shortens the travel time to important Hindu and Buddhist pilgrimage sites in the region and was expected to bolster the economy of the region by playing a key role in the growing Sino-Indian trade. However, trade is limited to specific types of goods and to specific days of the week.

It is also one of the five officially agreed- to Border Personnel Meeting points between the Indian Army and the People's Liberation Army of China for regular consultations and interactions between the two armies to improve relations and address differences.

In 1994, the relationship between India and China was peaceful and unthreatening. But then these remote places of India were still developing. The Border Roads Organisation (BRO) maintains and builds roads in all these strategic locations and ensures that the roads are serviceable at all times. However, during winter because of the heavy snow sometimes the roads get blocked. In summer the temperature hardly crosses 15° Celsius because of the high-altitude terrain. All the service personnel and families travelling from other states and regions, stay in Gangtok transit camp for at least a day or two to get acclimatised and then proceed higher towards the pass.

Some of these areas are named based on the milestones, say 5 Mile, and Ajit stayed in a location close to the ridgeline. Because of the severe weather condition, families were allowed only for a month in summer. Ajit, being a sincere follower of protocol, decided to take me to his location and I happily agreed, not aware of the hardships I was going to face.

It was during this Sikkim tenure that I was expecting my second child and it seemed well planned because Ajit wanted to have two kids and not stop with one! He felt that a single child would feel lonely and therefore for the sake of Nirbhay, I was carrying our second son, Akshay, when Ajit was posted in Sikkim.

Ajit had come down on his annual leave coinciding with my delivery date. I had almost completed my full term of pregnancy and we celebrated our wedding anniversary on 11th October 1994. I thought if I deliver on the 11th, it would be a double celebration, our wedding anniversary and the birthday of our second child. However, nothing eventful happened — after having a sumptuous wedding anniversary dinner we retired to bed. The next day I was feeling uneasy and at the same time anxious about my delivery, hoping for a baby girl. That afternoon I was busy with Nirbhay, feeding him his lunch after which Ajit and I had our lunch. After some time, while I was doing my chores, I realised that my amniotic sac had broken and water was trickling out. This was something serious. Ajit, my mom and I immediately rushed to the hospital, while my dad and Nirbhay stayed at home. I was panicking, while Ajit was cool and composed, as he drove us to the hospital. Once in the hospital, the nurses took good care of me, following the instruction from the gynaecologist, they induced my labour and lo behold a baby boy was born, after a due struggle! Yes, I was disappointed that it was not a baby girl! After the naming ceremony of our second child, Akshay, Ajit returned to Sikkim.

In April 1995, since the boys were too small for me to handle and travel alone, Ajit came down to Chennai on casual leave and helped me to prepare for the journey and one month of stay in Sikkim. We boarded the train to Siliguri, West Bengal. We couldn't take a flight, because the flight tickets were too expensive for our pockets and also, we were entitled to free travel from home to his place of work once in two years. We availed it, after a lot of paperwork and formalities. I learnt an important lesson while booking these tickets: nothing comes easily in life. I had to meet the concerned railway authority, who then checked for availability and finally approved the ticket, after which I had to go to the counter and wait for my turn, for the ticket to be issued. Quite a laborious task indeed to get entitled tickets.

Ajit with Nirbhay at Changu Lake

For the 72 hours of the train journey, I had diligently packed a steel flask with hot water, Cerelac for Akshay, fruits like banana and apple for Nirbhay, some packets of biscuits, *poori* and *sookhi aloo*, *idli* tossed in gunpowder (chutney powder as it is famously called), tamarind rice, and little curd rice which had to be finished within 12 hours of our journey. While we were heading East, vendors were seen selling only eggs and *murmura* (puffed rice). I was happy that I had well-stocked my snacks. Ajit would go with the steel flask, once every four hours, to fill hot water from the pantry in the train. Slowly Ajit evolved to be a very caring father, played with the boys in the train and managed to keep them engaged for the three full days of travel. Once we reached Siliguri, a military convoy took us to Gangtok transit camp. We stayed there for a couple of days and got acclimatized and proceeded higher up where Ajit was stationed.

Because of the tough terrain, inclement weather and security considerations, the movement of vehicles would stop by 3 pm. Wherever we had to travel, we had to plan in such a way that we start at 4.30 am to maximise the daylight hours.

Finally, we were at Ajit's location. It was a luxurious one room with attached bathroom. Though it was mid-April there was a hailstorm and I was advised to keep the kids and my feet warm. This was the first time I had experienced 'high altitude breathing', which was very heavy and we could hardly walk. So, for one month we stayed in that room, facing the majestic Kanchenjunga. The sun would rise at 4 am and by 10 am the sky would be packed for a heavy downpour. The view was very scenic and magnificent in all aspects. We had a 'Bukhari', a conventional room heater, to keep us warm at night, with temperatures dipping to 5 degrees and below. There was no electricity after 7 pm, so we would rely on two or three emergency lamps for light.

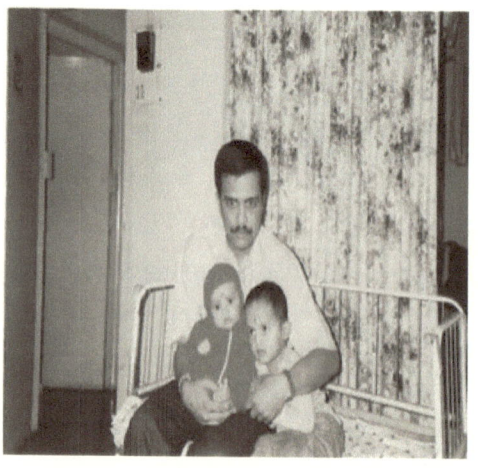

Ajit the doting father of two kids

While we were staying in the room, Ajit would freshen up, offer his prayer, which was the powerful *Gayatri mantra*, have his breakfast, which generally was eggs and toast and go to work sharp at 8 am. He would be back by 2 pm if he had no meeting or inspection of any sort. As days passed,the monotony of staying in a single room slowly started seeping in. Because of the high altitude, there would be hardly any fresh vegetable on the menu, the meals would always contain potatoes, different types of dal, rice and roti. Having travelled so far to spend quality time with Ajit I was disappointed to have a mundane routine. I must admit, in that location, I was given the best room, the facility of a kerosene stove, to provide hot water in the toilet 24/7, which was luxury in a place faced with water scarcity, and no cooking, since the mess catered to us. So, one day when I inadvertently told Ajit that I was getting bored, he immediately gave me that "cold stare" and said, "I have not brought you here for a holiday, but to show you the hardships which the jawans face." In spite of the challenges, they continue to perform their duties which is commendable! The summer stay in 17 Mile, was an eye opener for me and never again had I complained to Ajit about lack of facilities and resources.

During our stay, Ajit took us to the Nathu La pass. On that day they had the traditional "*Dak* Exchange". *Dak* is a Hindi word which means post. The "*Dak* exchange" ceremony is when the Indian soldier exchanges post and information with his Chinese counterpart, across the border. Ajit who was with me told me to keep watching the post, while he quietly went to do the "*Dak* exchange". Ajit had quickly worn a white jacket, and performed the exclusive ceremony of the Nathu La Brigade. Military traditions have their own beauty and significance.

Therefore, my brief stay with Ajit while he was a Brigade Major at the Nathu La pass was an eye-opener and this exposure was deliberately given with the ulterior motive of making me understand the conditions that the army operates in treacherous terrains. From then on, I could thoroughly understand all the hardships our jawans and officers go through, in protecting our country so that we can sleep in peace.

So, after overcoming few smaller hurdles Ajit and I, along with our boys were back in Chennai, and I had my whole family lauding me for all the effort I had taken to travel with two kids all the way to Sikkim. Something interesting happened, during my stay in Sikkim, I was spending my time reading books and newspapers. I got to know that admissions were open in Annamalai University for their long-distance courses, Bachelor of Education (B.Ed.) being one of them. I just shared my interest in pursuing B.Ed. with Ajit and showed him, the paper cutting of that advertisement. When we arrived at Chennai, the very next day Ajit informed me that he was going out for some work and would be back soon. When he came back, he had the application for the B.Ed. course, from Annamalai University in his hand.

At every stage of our life, Ajit was slowly and steadily empowering me to be an independent and a self-reliant mother of two!

FIRST BORDER PERSONNEL MEETING AT 63 MOUNTAIN BRIGADE

I was then the Brigade Commander at the Nathu La pass when Ajit was the Brigade Major. On 20th September 1995 the very first Border Personnel Meeting was held with the Chinese delegation, headed by Senior Colonel Wang Pingwen. After that, Ajit and I were discussing how this needs to be kept alive. Ajit then suggested that we should put a suitable plaque at the meeting site to keep the dialogue going.

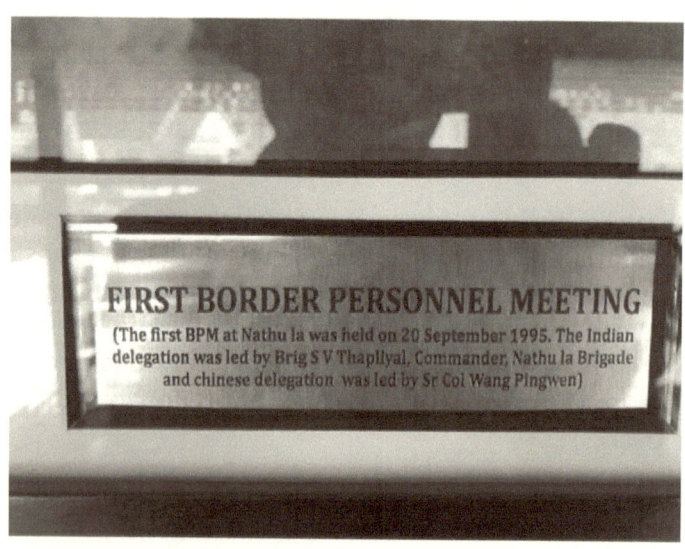

FIRST BORDER PERSONNEL MEETING
(The first BPM at Nathu la was held on 20 September 1995. The Indian delegation was led by Brig S V Thapliyal, Commander, Nathu la Brigade and chinese delegation was led by Sr Col Wang Pingwen)

I liked the idea and spoke to Gen Seth, the GOC, who approved the idea. Ajit then got the plaque made and it was put up at the meeting site. This spot now has become a tourist attraction at Nathu La, Sikkim.

Maj Gen S V Thapliyal (Retd)

Chapter 12

MS Branch - Army Headquarters: South Block, New Delhi

Military Secretary's office is the branch of Indian Army, that controls the appointments of all officers except the Medical Corps. It is also responsible for the promotions, postings, tenures and for the grant of honours and awards in the Indian Army.

This branch, generally called the MS branch, is headed by a senior officer of the Lieutenant General's rank and the Military Secretary is one of the six principal staff officers to the Chief of Army Staff (COAS).

Ajit with Nirbhay and Akshay enticed by the snake charmer in Udaipur, Rajasthan

After Ajit's tenure in Sikkim, he was posted at the MS Branch at Army Headquarters in New Delhi. We were all happy that we were moving to the capital, and that too after the two long years of separation from June 1994 to May '96. Ultimately, we were living together under one roof.

But then living in a big city had its own challenges. The Delhi cantonment catered to the officers from Delhi Area and its surrounding units. The army quarters were inadequate to cater for those officers posted at the Army Headquarters. Therefore, the Ministry of Defence had allotted many quarters amidst civilian areas too to meet the demands of defence personnel. The sanction of quarters was given based on seniority and the waiting list. We were given temporary accommodation in the Asian Games Village Campus, Lodhi Road. These accommodations were built for players who participated in the Asian Games of 1982, more like studio apartments with one bedroom, a small kitchenette and a living room. On conclusion of the games, these were allotted as temporary accommodation to Central Government officers' including those from Defence Services.

It took us almost three months to get our semi-permanent accommodation at Kaka Nagar, New Delhi. The reason we accepted this accommodation was that our kids were small and not going to any school in particular. So logistically it suited Ajit since it was close to his office in South block.

For the first time, I was waiting for the joy of the weekend. I longed for it! Otherwise, in the Army, it is always working 24/7, no public holiday or weekend culture! I was amazed by the fact that there was something called "office hours", and Ajit, at last, was doing a "9 to 5" desk job!

With two naughty brats to take care, my hands were full. However, Ajit was extremely supportive and would take the kids out to the park, while I was busy preparing for my B.Ed., written exam. He requested his parents to also come and stay with us for a few weeks. In the meantime, my brother-in-law, Col Arun Bhandarkar, also got posted to Delhi.

Ajit's father was a man of few words, he hardly spoke but was very fond of reading the newspaper. Both my parents-in-law had a strict routine, breakfast by 8 am, mid-day snack at 10.30, lunch at 1 pm, evening tea at 4.30, walk between 5 to 6 pm and finally dinner at 8.

One day when Ajit's father was having his mid-day snack, with a cup of sugarless tea, he could not keep his teacup on the table. His fingers couldn't respond and his left hand became stiff. Ajit immediately rushed him to the hospital, and after examining, the doctor confirmed that it was a nervous breakdown which is quite common amongst the elderly. After consulting the neurosurgeon, he suggested that my father-in-law should undergo a small surgery, for which both Ajit and his brother, Arun agreed.

During the hospitalisation of Senior Bhandarkar, the Bhandarkar brothers, Arun and Ajit, took good care of their father. Ajit would carry his father's lunch and some light food since Ajit's father was diabetic and preferred to have South Indian dishes like *idli* and *dosa*. After a week, my father-in-law was discharged and back home. Ajit was stressed and anxious about his dad's health, but did not express it. Finally, Ajit counselled him about his food habits and advised his dad to have light food and avoid fried snacks which he was very fond of. Ajit always intervened to dissuade his father from taking a cup of ice cream or gulab Jamun offered to him by a host. It was a tough job for my mother-in-law to keep control of my father-in-law's dietary requirement; Ajit's intervention made it easy for his mother.

Then came my B.Ed. exam and I was informed about the examination centre and the time table. Those days we didn't have Swiggy or Zomato for food delivery. So, I would cook simple meals and sit down to prepare for the exams. Ajit would be very accommodative and also would see to it that the kids also had their food on time. He too became a very good father to his little darlings, balancing his high-profile role with his family requirements. On weekends, Ajit would drop and pick me from the examination centre and on working days we managed to hire a cab.

Therefore, the Delhi tenure definitely brought us all close to one another as a family and before the extreme winter would set in, my in-laws returned to Bangalore, where they lived with their second son Anil and his family.

Ajit, hardly visible, standing at the back with his Sainik School Bijapur school mates

Birthday celebrations, get togethers and wedding anniversary parties, you name it, Ajit was a good host. His most favourite job, before the party, would be to wipe all the glasses and keep them spotless for the guests, and check on the menu. He would ensure that we get a few things like ice creams or confectioneries from the supermarket for dessert, to reduce my burden in the kitchen.

Weekends in the winter would be a long walk from our home to India gate, basking in the warmth of the Sun; Nirbhay on his small cycle and Akshay on Ajit's shoulders, while I tried to keep pace with them. Though this posting was administrative, Ajit did maintain his fitness regime: his morning runs in Lodhi garden and evening exercises at home kept him fit as a fiddle. He also suggested that I take time for myself and go for a walk in the evening.

The best part about Ajit's Delhi posting was that many of his school mates, course mates and relatives who were making their journey to J&K and were returning from there would halt at our place and then proceed onwards. In this process, I got to meet many of them and got myself acquainted with them. I am in touch with some of them to this date.

Ajit very subtly showed me that the organisation he was serving, is a family and I gradually became very out going and friendly. On introspection, my personality, as a protected child slowly changed into a confident mother and an independent *fauji's* wife.

CHAPTER 13

YUDHA KRITYA NISCHAYA: THE ARMY WAR COLLEGE; WHERE THE TIGERS EARNED THEIR STRIPES: AJIT'S PROFESSIONAL ALMA MATER

"War is only a continuance of the state policy by other means. But war must always serve as the larger ends of policy and not become the end in itself. The statesmen's objective must always be the betterment of the state as a result of war and not the defeat and destruction of the enemy."

– Kautilya

Ancient Art of Military Warfare

The Epics Ramayana and Mahabharat not only teach us moral life lessons but also theories of warfare and sophisticated weaponry; the *Chakravyuha*, used in the *Kurukshetra* War is one such example.

The first indication of institutionalised strategic thinking dates back to Kautilya's Arthashastra in 200 BC. The military strategies and its doctrines are relevant even today. The great emperor, Chandragupta Maurya, had a huge integrated force that consisted of all arms. The Battle of Haldigatti in 1576 between Maharaja Rana Pratap and Emperor Akbar is one among the many famous battles in Indian history which exemplified the use of tactics and operational art in the war stratagems. Tipu Sultan's rocket launchers against the British Army and the Sikh ruler, Maharaja Ranjit Singh's use of the modern army to delay the British colonization of Punjab are examples in our modern history. We also had Rani Jhansi, Rani

Abbakka from the Chowta Dynasty and Kittur Rani Chennamma, to name a few of our women warriors of the past.

Indian troopers, even as a part of British Indian Army, had fought in almost all major wars of the 18th and 19th centuries. In the 20th century, they showed their mettle in the two World Wars. Post-Independence the wars against Pakistan in 1947/48, 1965, 1971 and the Kargil War in 1999 clearly showed the acuity of the armed forces and their military acumen.

The strategy, operational art and tactics shared at the Army War College is the distilled wisdom of centuries of combat by great captains of war across the world. It is an amalgamation of the rich experiences and the lessons learnt from great wars from the ancient times to the present engagements by armies in the 21st century. The spirit of training at the Army War College is also influenced by the greats like Prithviraj Chauhan, Chatrapati Shivaji, Raja Raja Chola and other accomplished military leaders.

Post-Independence Era and Modernisation

The 1962 Sino-Indian conflict and 1965 Indo-Pak war highlighted the indispensability of an integrated all arms battle concept. Military victory was contingent on a commander achieving congruence and synergy of all arms and services. The stage was thus set for the merger of the Senior Officers' and Tactical Wings of the Infantry School in April 1971 to raise the College of Combat. Along the years the curriculum and the approach to warfare has changed manifolds, so therefore, the College of Combat was re-christened as Army War College on 1st Jan 2003.

"College of Combat" as it was called earlier, Ajit did his Junior Command Course 53 from August '88 to Nov '88.

His report reads:

"A very intelligent officer who is meticulous, keen and confident. His expression, both written and verbal is clear and effective. He exercises effective control over the group. His knowledge of his own arm is above average and that of other arms and services is high average. His analysis of tactical problems is logical. His understanding of battle procedures and battle drills is clear. He has a keen eye for ground and can produce sound solutions to tactical problems. His comprehension of pattern of battle at unit and sub-unit level is clear. His overall performance on the course had been high average."

Col Suresh Chandra   **Brig Chandiramani**

Senior Instructor   **Commander JC Wing**

At every stage of promotion, an Army officer has to qualify a few mandatory courses to pick up his rank. While the JC is for a Captain to become a Major and assume the responsibility of Company Commanders; Senior Command (SC) is for a Major to assume the appointment of Commanding Officer of an Infantry Battalion or equivalent units in other arms and services.

Ajit attended the SC 80th batch, which started on 8th December '97 and concluded on 7th March '98. Since we had two small kids who had not yet started their formal schooling, Ajit decided to take all of us along with him. We had a Maruti Omni, which was filled with two trunks of household essentials along with two suitcases of clothes needed for the duration of the course and we drove down from New Delhi to Indore. Ajit, being an expert in navigation and map reading, planned in such a that way we halted in the evening in some army guest room.

Those were the days when we didn't have a smartphone with Google Maps to navigate or an induction stove for cooking! Good old tough days! The hard copy of the map was on the dashboard for reference and I took on the role of a navigator, warning Ajit of any speed breakers or diversions, while Akshay and Nirbhay were at the back seat either sleeping or looking out of the window. Some places the roads were appallingly rutted but however, we managed the drive and reached Mhow, on time to report for the course. The best part about Mhow is you will get any item on hire, be it a mini-refrigerator, gas stove, desert coolers, etc which was so convenient for the part-time denizens of this historic town. The icing on the cake was that eight of his NDA course mates were also doing the course. Officers with families were allotted the usual one-bedroom studio apartment, with a kitchenette and a drawing-room.

Ajit with Nirbhay and Akshay in Amber Fort, Jaipur, Rajasthan

Only a few officers had got their families, since many of them had kids who were in school. Families who had children of a similar age as ours were Col C Ramchandran and Col Pradeep Katoch; both retired as Colonels who were majors then. Some of the other officers' families joined the officer for a week or so during the Christmas break. Therefore, Col Ramachandran's and Col Katoch's family became very close to us and we would go out together, whether it was for purchasing grocery or shopping in the Mhow market or even attending the AWWA meet for the ladies when the officers were busy with their classes. All our kids were enrolled in the nearby Shaurya Primary School, Mhow for a short period.

Life in Mhow was hectic for all officers and their families. After classes, the weekly movie in the open-air theatre and sometimes picnic during weekends to Mhow fort or shopping at the *'kapada'* market at Indore did provide us the occasional respite. Finally, after the course got over, we once again drove back to Delhi but this time we drove via Jaipur, Rajasthan, covered the Amber Fort, the City

Palace and the Hawa Mahal. After Jaipur we stopped at Agra, visited one of the seven wonders of the World – the Taj Mahal – and reached our home at Kaka Nagar, safe and sound!

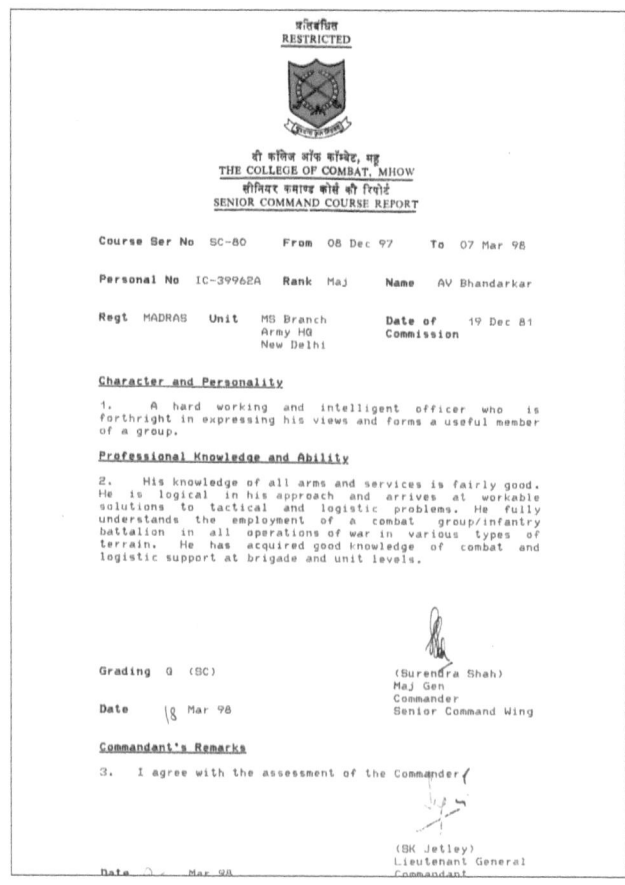

SC Report

Note: Presently, Mhow has been renamed as Ambedkar Nagar, after Bharat Ratna, Dr Bhim Rao Ambedkar. His father was a Subedar Major, a Viceroy commissioned officer of the British Indian Army's Mahar Regiment. A Buddhist Stupa has been built at a spot where his father's quarter was located. The Agra-Mumbai road runs very close to the memorial; close to it also lies the temple, gurudwara and mosque of the Infantry School, Mhow.

CHAPTER 14

DRIDHATA AUR VIRATA: 25 RR

हतो वा प्राप्स्यसि स्वर्ग जित्वा वा भोक्ष्यसे महीम् ।

If thou in battle slain, though will surely attain heaven

- *Gita/II/37*

Our family life of four, my career and Ajit's rank were picking up pace. Nirbhay was studying at Army Public School, Dhaula Kuan, and Akshay was at Raj Kumari Amrit Kaur Study Centre attached to Lady Irwin College, while I started working in Green Field School, Safdarjung Enclave, as a primary school teacher. Ajit was a highly supportive husband. Our day would start at 4.30 am, while I would get all the lunch boxes, snack break for kids and the breakfast ready, Ajit would go for a run and be back by 6 am. I would wake up the kids by 6 am to get them ready for school. While I got ready by 6.45 am, Ajit would drop me at the bus stop where the school bus would come to pick me up by 6.55 am.

My colleagues would be surprised that by 6 am I would have already prepared two major meals i.e., breakfast and lunch along with the snack break boxes for the kids which generally consisted a fruit and some biscuits. The secret of this feat was that all the preparation would have been done the previous day. Like in all south Indian homes, the *idli* or *dosa* batter would always be there in the fridge. Once Ajit was back from dropping me at the bus stop, he would ensure that the kids had their breakfast that was kept on the table and board them on their respective army trucks. Yes, in the Army, for the safety and security of the children, many modified three-tonners would pick up and drop kids from their homes. Of course, we did have a good caretaker and housekeeper who helped us with the cooking and babysitting.

As a teacher too, I had to count the marks the kids scored in their various subjects, prepare the report card manually by entering the marks and also complete lesson plans for the month. Back then, neither did we have the excel sheet to do the grand total of marks at the click of the mouse, nor did we have the computerised report card. Ajit would assist me in the counting of marks and sometimes dictate the marks of the students so that I could write faster. We would have candid conversations upon the performance of the students, wherein Ajit would try to understand why that particular student had not done well, which thoroughly showed his compassionate nature!

We all knew that it was not a permanent posting, and anticipated the next posting order a couple of months before the third or the second year of the tenure whichever applicable. So, this posting was for three years and we had completed two years and eight months. One day the discussion on the dining table was of posting, where could it be? Having known Ajit for his preparedness, he told me that he will be picking up his rank after completing the Senior Command and will be posted as a second-in-command of a Rashtriya Rifles (RR) battalion!

We had lost Maj Dhaiya, 18 MADRAS when he was deputed with 8 RR Madras in 1997, so I was apprehensive about the RR posting. However Ajit immediately retorted stating that if all of them fear going to RR, then who would be at the borders protecting the country, which was the primary role of the defence personnel!

I listened to him carefully and suggested that since he was in the branch which was handling important appointments, couldn't he make any changes? For which he replied that it will be against the ethos of his job! After this conversation we never again discussed his prospective transfers, and as per Ajit's presumption he got his posting as the 2 IC of 25 RR battalion. After which, we decided to opt for the separated family accommodation at Delhi so that we don't disrupt the children's schooling and also because it was closer to J&K.

Rashtriya Rifles is the Indian Army's elite counter-insurgency force in Jammu and Kashmir, which is prone to militancy and terrorist infiltration. The RR unit was raised in 1990 by the then Army Chief, Gen VN Sharma with Lt Gen PC Monkotia as its first Director-General. Presently, the force comprises of 65 battalions, divided into five companies: Romeo Force, Delta Force, Victor Force, Kilo and Uniform Force. Half of its strength comes from the Army's Infantry units and the other half from the other Corps of the Indian Army.

All officers, JCOs, and ORs prior to joining their respective units must undergo four weeks of rigorous pre-induction training. It is followed by two weeks of 'on the job training' and a periodic refresher training under sector arrangement. All these battalions are not just made of sweat and blood but also have a heart and a soul. These units conduct community goodwill activities and take part in the grassroots development of local civilians through a program called Sadbhavana.

In 2018, the RR celebrated its silver jubilee, successfully having neutralised over 16,368 terrorists, which include 8, 522 slained, 6737 apprehended and 1,109 surrendered.

On 5th June 1998 after being promoted to the rank of Lt Col, Ajit reported to 25 RR as a second in command which was based in Doda, later on moved to Surankot, Poonch district. The only means of sharing information and being in touch with each other during his tenure would be through calls from the Army exchange and letters which we would write frequently.

In the summer of 1999, we had vacated our quarters and had moved to Bangalore. I stayed in Bangalore for two months with my parents-in-law and then we were back in Delhi. It was during this time that the Kargil War had broken out. We, in Bangalore, were watching the news very closely. We had 527 soldiers killed in action during the Op Vijay operation and on 26th July, India emerged triumphant over Pakistan. Ajit was communicating through letters and I was extremely busy in grooming and engaging my two brats.

At 25 RR - Ajit standing second from the right

In the meantime, Col GK Rao, a very close friend and the course cum squadron mate of Ajit, suggested that we, two kids and I, stay in their army quarters, along with his family, till the time Ajit was allotted an accommodation. If not, I had to stay in Bangalore and the kids would miss school. So, we moved into Col GK Rao's accommodation, where his wife Sucharita and I bonded like sisters, and their kids Varun and Yamini enjoyed the company of Akshay and Nirbhay whilst I continued to go to work.

Now, with regards to the Separated Family Accommodation, it was a brand new one, which meant a slightly longer wait to complete the handing over and taking over procedures. After a long wait, we were allotted the accommodation at Manekshaw Marg, Delhi Cantt. At the same time, the information we were getting from the J&K sector after the Kargil Vijay Diwas was not pleasant. We were losing jawans and officers very frequently to terrorist attacks since there were many terrorists still holed up in the valley.

Ajit as a second in command of 25 RR was meticulously chalking out his strategies for his troops and studying terrorist activity so that the *paltan* could eliminate the terrorists who were trying to instil fear in the minds of the people and disrupt peace in the beautiful valley.

106

Since 25 RR was a newly raised unit, Ajit had travelled to Delhi to purchase some basic sports equipment for troops to ensure their physical fitness as they worked in high altitudes. Team games like volley ball, basketball, etc. were played in the evening and also inter-company matches brought lot of cheer and josh in troops. So, to sum up, Ajit was always working for the welfare of his troops and simultaneously scheming to neutralise terrorists in his area.

As a mother, to make the boys sensitive to the situation their dad was facing, I made my kids write short sentences, draw pictures

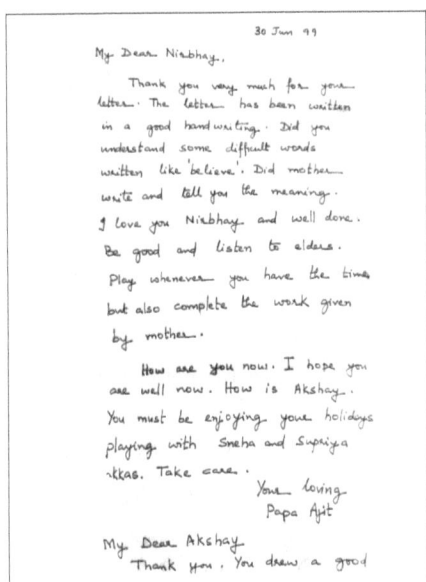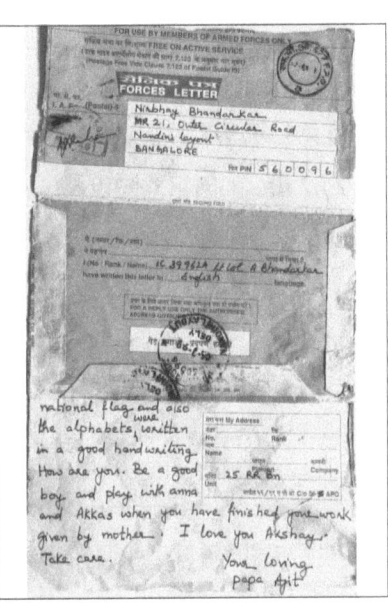

The letter from Ajit to Nirbhay and Akshay

and send cute greeting cards to their dad to lift his spirits. On 9th August 1999 he came over to Delhi, on a short trip to wish me on my birthday, though we didn't celebrate it that year, because of all the terrible loss of lives and casualties. The main purpose of the visit was to help me settle down in the new accommodation, he also checked for all the facilities available in the campus, ensured that the school bus facilities both for me and the kids were fixed, so on and so forth.

Finally, we had settled in the new apartment quarters at the SFA. I generally spoke to Ajit through the Army Exchange calls, finally one day, he spoke about the situation on ground in the forward area, stating that it was infested with terrorists and anti-nationals. So, to keep him motivated, we would write letters regularly. On Friday, the 29th October 1999, in the evening I had a telephonic conversation with him, during which he discussed about the garage alloted to us, the condition of the buses and trucks to our schools and enquired about our boys Nirbhay and Akshay. After I reported about the kids' activities and their routine to Ajit, he hung up fully convinced about their progress and development. He also spoke a couple of motivating statements, stating that I was a strong and smart woman and told me not to get demotivated by his absence.

On the next day, 30th October, Saturday, I was busy designing my teaching aid for my class, the boys were sitting in the living room watching their favourite cartoon 'Tom and Jerry'; suddenly, the doorbell rang and I went to open the door. Many of my colleagues, from Green Fields School, their husbands and one officer from the Delhi Sub Area had accompanied them. I warmly welcomed them and asked them for tea or coffee, but they all refused. They spoke to the kids and also enquired about me and finally, one officer told me that Ajit was no more! I was dumb struck. Initially, for some time, I did not believe the news, I said, there must be some miscommunication because I had spoken to him the previous day! I told the officer in charge that unless I get to hear from an officer in 25 RR, I will not believe him.

I can never forget that terrible day, Nirbhay was seven and Akshay was five years old, unaware of what was happening around them and they were innocently oblivious to the whole situation. That night I was sleeping alone, hoping that someone would call me to say that it was all a mistake of identity. The next few days were remorseful and melancholic. All the unit officers, his course mates, school friends, neighbours and friends came over to convey their condolences while I was yet to realise the intensity of the calamity.

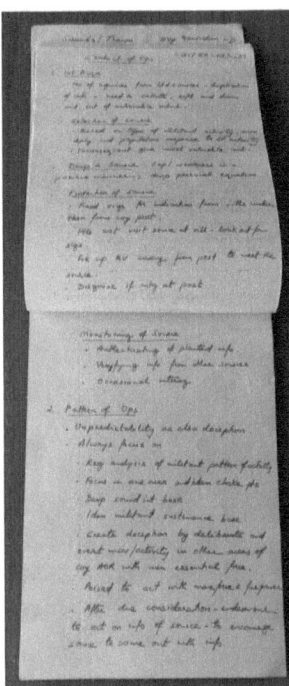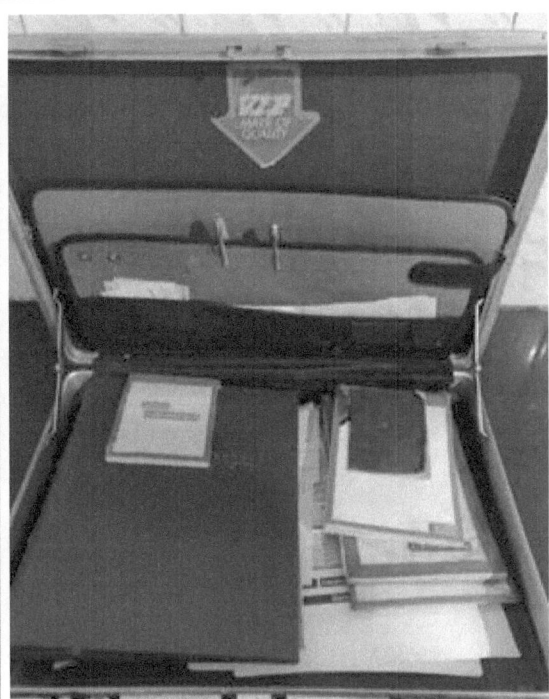

*Ajit's Briefcase containing his diaries, log books and notes meticulously
maintained before his final days*

After the unfortunate event of losing my dear husband on the
battlefield there was a perpetual feeling of hollowness and a dull ache
in my heart. I tried to put on a brave face in front of others; always
dreading the gush of sorrowful memories which came in torrents. I
tried to concentrate on the rejoiceful memories and immersed in the
stupendous task of bringing up my two sons, whom I would have to
nurture without the presence and support of their father.

A couple of days back while exploring my attic I came across Ajit's
briefcase which he was using till the final days of his life. I perhaps
had kept it out of view always fearing the avalanche of agonising
memories that could inundate me? Finally I took the plunge and
decided to open it and lo behold, found his log books, diaries, notes
and plans!! He had the names of terrorists of the area, their family
tree, names of Panchayat members, prominent personalities and
a plethora of information of the area. It was beyond doubt that he

had a close rapport with the locals and his intelligence collection and collation system was well streamlined. Later, I came to know from his friends that without intelligence of the area of operation it only ends in fatigue of troops as nothing much can be achieved. I also came to know that the Second-in-Command is assigned with the responsibility of collation of information and derive meaningful actionable intelligence out of it.

Ajit's painstaking effort was quite evident in his note books and logs. He had always been exacting on his team mates in the professional field and spared no one; not even himself, in preparing for operational tasks. Ajit was passionate professionally and took immense pride in accomplishment of his mission.

SECOND IN COMMAND: 25 RR- LT COL AJIT V BHANDARKAR

When I got posted to 25 RR in 1999 as a young Capt, Lt Col Ajit Bhandarkar (Ajit Sir) was the Second–in–Command (2IC). There are volumes of goodness to be spoken and written about Ajit Sir which I am sure hundreds of people have narrated and shared. Today I will narrate one specific military operation in which we both had roles and I watched with awe the personal involvement and leadership of Ajit Sir. I have total respect for him as a military leader and as a genuine human being.

In June 1999, 25 RR had moved from their location Bhaderwah (Doda Distt, J&K) to Surankote (Poonch Distt, J&K) in the thick of OP VIJAY. Life was tough, the battalion was continuously carrying out military operations in our area as we were new and we wanted to know area of responsibility at the earliest. Like all RR outfits, our battalion was in action from the very first week at the new location.

On 08 July 1999, the Battalion HQ received information from Army Intelligence about 4-5 terrorists in an area called Darra Sangla. I was the STF Commander (Commando/Ghatak Platoon Commander as they are known now) and also the officiating Adjutant. The CO, Col Ramachandran (Ramu Sir) called Ajit Sir and me to give the information from Army Intelligence just before lunchtime. Ajit Sir told me to make the initial plan and discuss with him during lunch. Over lunch he heard me patiently and gave some more valuable inputs. He had such experience of operating in J&K that his advice was to be taken seriously; he had a spoken reputation of being a natural leader. The same evening, CO and 2IC heard me brief the troops going for action. Ramu Sir had detailed his own crack team to go along with my team under my command. Ajit Sir assured me that if more troops are required then he along with his Quick Reaction Team would back me up. My team got launched into the operations at nightfall; we walked whole night to reach the area of terrorists. The next morning (09 July 1999 at about 0630 hrs), we spotted two terrorists and after about an hour's gun fight we killed both. I informed Ajit Sir on Radio Set who was with CO about our success, he told me not to look for the other terrorists and immediately fall back with the bodies to the local Police Station. My team collected a few locals to carry the dead bodies to the Police Station; we started moving back at about 0800 hrs. On our way back, we encountered heavy volume of fire from the nearby village and the gun battle started which lasted for 6-7 hours till 1500 hrs in the afternoon. I contacted Ajit Sir on Radio Set to inform him of the second encounter, he had immediately started with his team to our area of action. When Ajit Sir met me at about 1500 hrs, he had already cleared the terrorists firing from the nearby village (we later came to know that four terrorists were dead in the second encounter of the day). He shook hands with me to congratulate me for having a long hard day and a few kills too but saw my dress covered in blood. On his enquiry, one team member who was near me revealed that I had got bullet injuries at 0900 hrs in the morning. I told Ajit Sir that I didn't want my team to panic because of my injury so I didn't inform him/CO on Radio or to the team and moreover I was able to use my AK Rifle with one hand so I didn't feel necessary to inform about my injury. Ajit Sir got wild at my naive reasoning because he knew from his vast experience that loss of blood can be dangerous to life. I saw him very closely working towards my evacuation and medical treatment; he called for helicopter evacuation which couldn't land even after hovering over us due to some technical issue. Ajit

Sir was pissed off with the helicopter guys because I had by then started feeling weaker due to more blood loss. I heard him saying to someone at higher HQ over Radio Set -- "Sir, if something happens to this officer, I would hold all you guys sitting on the chairs at HQ responsible. Do anything to arrange a helicopter for evacuation." Again another helicopter hovered above us but couldn't land. Then Ajit Sir just called a JCO/ NCO and told him to take a few boys to get a charpoy prepared for me to be carried. In next five minutes, I found myself strapped to a charpoy, saline bottle tied on a bamboo above me and being carried by 4-6 jawans. Ajit Sir walked for hours with me on charpoy until we reached Surankote Military Hospital at around midnight. There I was operated and days later I got stabilized. This is how Ajit Sir had total involvement with his officers and men, he was instrumental in saving my life by taking the timely decision to evacuate me on foot and not to wait further.

Months later after my sick leave in end October 1999, I was in Military Hospital at Guwahati for my medical review. When I was listening to AIR News on radio after dinner, I came to know of Ajit Sir being martyred in a military operations in Surankote. It was a sad day, may God bless his soul.

Our first child was born in February 2000. I had decided to name him/ her after Ajit Sir, so I named my daughter "Ojeeta". After I recovered and joined office, I visited Ajit sir's family to show my gratitude to him.

It was a pleasure to connect with sir's family again. It's a proud moment for us that both his sons are towing the armed forces career of their brave father. We will always remember Ajit Sir, salute to the brave heart.

Col Shourav Nandy

16 Madras (Travancore)

MUSINGS OF OUR ASSOCIATION

Our battalion was located at Bhadarwah, a tehsil of Doda District in J&K. I, as a young captain had been part of 25 RR for about 16 months when Lt Col Ajit Bhandarkar joined us as our new Second in Command (2IC). Within next few days, he visited all company/ platoon posts and developed a firsthand understanding of terrain, insurgent network and intelligence set up our Battalion had created in its Area of Responsibility (AoR). He had an uncanny knack of connecting with people of all hues and colours. We the "youngsters" watched him both with awe and pride. His humility, straight-talk, fearlessness and solicitude simply qualified him to be a role model for my generation. Months later, as part of "Op Vijay" relocation, our Battalion was moved to Surankot.

Surankot, compared to Bhadarwah (Doda), was heavily infested with militants and our troops were encountering them more frequently. On one such occasion during July 1998, while my company was carrying out the task of 'Road Opening', we encountered few terrorists who were positioned to target the ammunition convoy moving to Poonch. During exchange of fire with terrorists at about 0430h in the wee hours and the resultant chase, we realized that we were outnumbered and that too in the thick of a deciduous forest. Visibility was low and reinforcement from company headquarters was not readily available. I think amygdala hijack or sixth sense takes over the decision making during such desperate and dangerous situations. Not even sure of his whereabouts, I made an SOS call to my 2IC on radio at about 0500 hours. He listened to me with rapt attention and very briefly inquired about my location. Meanwhile, we attempted to fire back at terrorists by changing locations, trying to create a façade of more strength than we actually had. At about 0730 hours, the radio-set by my side crackled and the "Good Samaritan" (Ajit sir) announced his arrival. Travelling time from Battalion headquarters to my location by fastest means was not less than two hours! As I briefed Ajit sir about the situation, his QRT cordoned the area and he assumed control of the operation. That day, we eliminated two hard-core terrorists without any injury to our boys.

Ajit sir apportioned full credit of the success to my team. It is hard to find such a breed of leaders even in the uniformed cadre. They say "Perfection" is a Godly business and we humans can never be perfect but Ajit sir came very close to being one. Two months after I left 25 RR, I learnt about the martyrdom of this braveheart. Our salutations and homage to the noble soul.

Col BS Poswal, SM

Army Air Defence

CHAPTER 15

SHAURYA CHAKRA: GALLANTRY AWARD

Nirbhay and Akshay were oblivious to what was happening around them. It was unanimously decided that the last rites of Ajit will be performed in Bangalore because his parents and all our family members were there. While all of Ajit's course mates and school mates from the Indian Air Force were tracking his mortal remains from Srinagar, J&K, where he was given the ceremonial send off and then on 1st November, in New Delhi again after the wreath-laying ceremony by dignitaries it was brought to Bangalore. The flight landed late night on 1st November, with the mortal remains of dear Ajit.

I had reached Bangalore on 31st October, via Indian Airlines flight with my two kids and went to my parents-in-law's house. Ajit's parents were old and totally shocked at this news. My mother was thoroughly depressed to know that Ajit was no more. Ajit's father was speechless and his mother – tears were rolling down her cheeks and she literally swooned. I was like a zombie who was only physically present but mentally cut off from real life. Both the boys were unmindful of what their future would be without the support of their dear "Papa", this is how they used to address Ajit. They would miss the toss and tumble, fling and swing, roll, romp and jostling which fathers sport with their children. Such instances were always exhilarating; never again such antics would ever be pulled off! Sitting on papa's shoulders, the evening walk with him, as he escorted them, when they rode their bicycles, and all the security of having their "papa" by their side, would now remain only in the inner labyrinth of their memories.

On the 2nd November, the pandit, who had performed our marriage, was there to do the final rituals of Ajit. The priest finally called both Nirbhay and Akshay to their "sleeping" dad and told them to give one last hug to his mortal remains. Akshay remained perplexed while Nirbhay vaguely knew something was amiss! According to the traditions, the sons i.e., Nirbhay and Akshay have the right to take part in the various ceremonies alternatively, any of his cousins could do it. I did not want my boys who hardly knew the seriousness of the issue to get into all the ceremonies. So, Ajit's first cousin, Arvind Bhandarkar, did the last rites and performed the pooja on the 13th day after the funeral, the *Vaikuntha Samradhana,* when the soul finally gets liberated, is put to rest.

The day of the funeral was a tough day, while all the family members, relatives and unit officers and his friends were there at MR21, Outer circular road, Nandhini Layout, Bangalore. The press photographers added to the crowd, they took pics, spoke to some of our relatives and followed the decorated three-tonner to the crematorium. So, with all military honours we had a grand farewell to the hero, 'veera yodha' and the martyr who gave his today for our tomorrow.

What had happened on the morning of 30th October, was that after attending the monthly, regular Commanding Officers conference at the sector headquarters, the COs were informed about some terrorists trying to enter the Indian territory in the Surankot area. The CO on his return to the unit, during his meeting with the unit officers, informed them about the inputs he had gathered about the terrorists trying to enter Indian soil. The CO discussed the details with his officers on the future course of action. Ajit, after discussion with the CO, volunteered to lead the QRT team. He made detailed plans, organised his team with manpower, weapons, ammunition. He briefed his team and also made logistic preparation and considered several contingency plans. The team was all set and proceeded on its mission, in the evening.

The operations commenced; Sepoy Binu, his buddy, was in front of him, and Hav Shiva Kumar another buddy was at the back, and he was behind the leading element of the QRT. It enabled him to exercise command and control over the entire team and he was giving the necessary instructions to the team. They were going for a search operation to eliminate five hardcore terrorists hiding in the border village. All of a sudden, Ajit spotted the terrorists escaping towards the *Nullah*. He started chasing them and shot dead one of them at point-blank range. The second terrorist was hiding behind the *Nullah* and was firing indiscriminately, injuring Ajit and also his buddy, Sepoy Binu grievously. While his buddy fell unconscious, he moved and lobbed a grenade at the terrorists. He then crawled forward and shot down one more terrorist. Thus, Ajit eliminated three hardcore terrorists, however, he also had taken a bullet on his forehead, just below his helmet and also got multiple splinter injuries on his right forearm. He displayed gallantry and bold leadership utterly unmindful of danger by leading from the front in the true spirit of his Regiment '*Swadharme Nidhanam Shreyah*: it is glory to die doing one's duty. The hero succumbed to his injuries.

To the left, Mrs Shakunthala Bhandarkar, Mother of Lt Col Ajit V Bhandarkar, SC, in the centre along with other army officers

The announcement letter for the Award

Keeping in mind the operation, and the supreme sacrifice made by Ajit, the President of India awarded him Shaurya Chakra, posthumously, a peace-time gallantry award on 11th August 2000. However, due to some inevitable reason, the ceremony was conducted on 12th October 2001. This was the first time in the history of 25 RR, a senior officer was killed in action and was awarded a gallantry award.

My mother-in-law, Mrs Vasudev Bhandarkar and I went to Delhi to receive the award. The day we had the ceremony, 12th October 2001, was very close to our wedding anniversary and was Akshay's birthday. On 11th October 2001, on our wedding anniversary day, we had a rehearsal of the ceremony. I couldn't stop sobbing, during the rehearsal, I had my liaison lady officer who came up and comforted me. Yes! no doubt, it was a proud moment for us, but it is a great personal loss to the family. We also had Ms Reena Muthanna, wife of Late Maj M Chinappa Muthanna, SC from the Sikh Light Infantry, who had come with us from Bangalore for the ceremony.

Exactly eleven years after our wedding, I was standing there alone, in front of all dignitaries, rehearsing for India's most coveted Defence Investiture Ceremony, for the presentation of Gallantry awards in the magnificent, Rashtrapathi Bhavan. The then President, His Excellency, KR Narayan awarded the medal to me, and while handing over the medal, he said: "We are proud of you". This sentence truly overwhelmed me, that too coming from the President of India boosted my morale.

CHAPTER 16

CHIP OFF THE OLD BLOCK: BOTH BOYS JOIN THE ARMED FORCES

Following the martyrdom of Ajit, we decided to relocate to Bangalore. I was told by some of Ajit's course mates that I could stay in the Army accommodation for two years, by then I should be able to settle down with a job for myself and a good school for the kids.

To cut a long story short, the boys were admitted in National Public School, Indira Nagar and I got a job in Army Public school. We were able to stabilise our work and routine during our stay at the Field Marshal Cariappa Officer's Colony, MG road. Thereafter, I decided to look for an apartment near Indira Nagar, so that it would be comfortable for the kids and I could travel to work at Army Public School.

Finally, we found a compact two-bedroom apartment near National Public school, which was also very close to HAL airport road, Bangalore. While years passed by and in no time, it was the right phase for my elder son, Nirbhay to choose his profession. Till grade 12, he was not sure of what he wanted to be, neither did I force him to take up any particular profession.

As a single parent, when my boys were in their teens, I enrolled myself in a counselling course, at Banjara Academy, so that I would first be able to handle my own brats. That was not it. I also joined the Art of Living courses and started my *Pranayam*, breathing exercises and *Sudarshan kriya*, which is still helping me lead a balanced life. Since, I had to take on the role of both the father and mother, I had equipped myself to be their friend, in all possible ways, whether it be watching a football match with the boys or be it flying in a microlight aircraft. As a counsellor, I had taken both my boys to premier institutions like the IIT Chennai, where my brother, Gopal Krishna Kamath graduated from, to the NDA where their dad graduated from and also escorted them to a few universities in the US and Canada with the help of my brother Narendra Kamath who lives in Canada. I had given them the opportunity and had empowered them to make their own choice and take decisions in their lives.

Family pic taken in Mussoorie, 1998

The parent unit of Ajit: 18 MADRAS (Mysore), always invited us for the integration day and so we would religiously attend the events hosted by the unit. One fine day, on our way back from one such event, Nirbhay said, *"Amma, I want to join the army".* Initially, I thought it as just a random statement made by him. But I was wrong. Nirbhay meant it and finally joined his dad's own unit, the Marauders 18 MADRAS (Mysore) after completing his graduation from Christ University and training in the prestigious OTA, Chennai.

Similarly, in Akshay's case too, he cleared the NDA written exam, after completing his 12th Board exam, and also the CET of Karnataka. At this juncture, we sought the help of a counsellor to help and enlighten Akshay about both the professions and its pros and cons. Akshay after attentively listening to the counsellor, who was one of Ajit's school mate, Capt. Patil, he decided to pursue his undergrads in Engineering from PES university, Bengaluru. While doing so, he would have better clarity about the career he chooses and would take an informed decision.

But then, trust me, every time the boys were in a dilemma and needed some guidance, I always felt the vacuum in our lives. How I wished Ajit, their father, was there for them to guide and talk about the choices they make in their lives. Never the less, I always felt that he is guiding us and protecting us from wherever he is.

Looking back at my life, many of Ajit's school mates, course mates and unit types have always helped me to tide over my storms with ease and perseverance. I am indebted to them and my boys both Nirbhay and Akshay fondly remember all of them and look at them as "God fathers" for any guidance and advice.

31st December 2020 happened to be Ajit's 60th birth anniversary, and I felt I had to celebrate his life and share his life's journey with the world so that it inspires the youth and motivate the millennials to look beyond the pay packet. Today, if I have to look at the Army as a profession, it has the toughest selection process and only 100 out of every lakh of the applicants make it to the services. I am happy that both my boys had the calibre and made Defence Services as their choice of career.

However, when Nirbhay made it to his Father's regiment, I had many people asking me, "Why did you send your son to the Army?" I would politely tell them that I did not send him, he got selected for the Army and it was his choice. I still remember the day, when Nirbhay was preparing for the SSB interview, as a mother I would ask him some GK and current affairs questions as and when time permitted. During one of such revision sessions with Nirbhay, I

*Left to right: Akshay Bhandarkar, Mrs Shakunthala A Bhandarkar
and Nirbhay Bhandarkar*

asked those "situational reaction questions" which was - "You have your board exam, and in the very morning, your mother has a heart attack what would you do?"; so, he very tactfully replied that he would give instructions to his brother to take his mother to the MH and get her admitted. He would then proceed to the examination centre to write his board exam. In this way, he saves his mother and also be able to attend to his exam. After completing his exam, he would then directly go to the hospital to meet his mother. All these precious moments of learning and observations in our life bonded us like never before!

Akshay, during his seventh semester got selected through the campus placements for the final round of interaction with the HR team of a MNC. During the interview, they asked him about his family, whether he was ready to relocate to any part of India and if he was ready to leave his mother alone. He promptly told them that his mother was independent enough and can hold the fort by herself. On that day, I felt so proud of him! We as a parent should learn to let go of our kids when they have to leave the so-called nest. Only then, will they learn to fly on their own and make a life for themselves! Akshay subsequently on completion of his studies joined the Indian Naval Academy, Ezhimala, Kerala, through the University Entry Scheme.

I guess the urge to join the Services was in their genes, both Nirbhay and Akshay followed their passion and, of course, their dad's footsteps!

The legacy goes on… Jai Hind!

Rest easy Ajit, you have given your all to the nation and we will, forever be indebted to you.

About the Author

Mrs Shakunthala Ajit Bhandarkar has done her Masters in English, BEd and a diploma in Counselling. Apart from this, she has also done a certification from Indian Institute of Management, Bangalore on 'Nurturing Creativity and Excellence' and has recently completed a course on 'Strategic Leadership for Schools in a Changing Environment' from IIM Ahmedabad. She has three decades of experience as an educator, mentor and counsellor. Currently, she is working as an Academic Consultant and an Instructional Designer for educational institutions.

She has been appreciated and honoured by various institutions like Infosys, BEML, PES University, New Horizon group of Institutions, etc. for her contribution in the field of education and women empowerment.

To commemorate the sixtieth birth anniversary of Lt Col Ajit V Bhandarkar, SC and their thirtieth wedding anniversary, Mrs Shakunthala Ajit Bhandarkar took the initiative to write this book which is her maiden effort. She stays in Bangalore and can be reached at shakunthala@ajitbhandarkar.com.

 **METAMORPHES**

THE TEN

In the month of January of 1978, 10 young boys of seventeen or so joined the National Defence Academy. Young and enthusiastic, having cracked one of the toughest entrance exams in the country, they were eager to face the world with a sparkle in their eye.

Three years of moulding and harsh, unforgettable training at NDA and thereafter at the Specific Army, Air Force and Naval Academies METAMORPHIZED them from boys to men. After passing out from the Academies as Indian Commissioned Officers with the coveted President's Commission, some wore blue and flew the skies, others in white scythed through our oceans but majority wore the Olive Green and went on to guard our borders.

Young Officers with nary a care in the world they were on top of their worlds, with strong convictions and commitment, doing good the only way they knew. They were embraced by their battalions, regiments, squadrons or ships which became their family, their home away from home. Each one chanced upon a lovely damsel who suddenly became the love of his life and a life partner. The better half, who actually made him better and complete, became the mother of his children and his cup of joy was full.

Alas one of them took on the mantle of his duties to the supreme level and was martyred.

THIS IS HIS STORY

Eight of the young boys, now veterans, teamed up with an objective to make this a better world. Priti, a senior Journalist and an empowered woman in her own right as also the better half of the ninth joined this band of eight. These nine teamed up with Shakunthala, the Veer Nari, to become –

THE TEN
[Priti | Shakunthala | Anuj | Ashwani | Nishant | Pankaj | Probal | Ram | Ramesh | Sudhir]

A singular aim to give back to society transformed the **TEN** to **METAMORPHES**; an idea, evolving into a series of initiatives to make this a better world for our Veterans, Veer Naris, the youth and the society at large, across borders.

An ode to **AJIT** *from* **METAMORPHES**

We will not let the world forget what you stood for

Take off your boots and rest Brother, you have done your bit...

Shakunthala has graciously offered the proceeds of this book to 'Metamorphes' to be used for the welfare of Veterans and Veer Naris.

Kudos Shakunthala for bringing Ajit alive... and Thank You

www.ingramcontent.com/pod-product-compliance
Lightning Source LLC
Chambersburg PA
CBHW031119180526
45160CB00002B/24